Late Quaternary Sea-Level Correlation and Applications

NATO ASI Series

Advanced Science Institutes Series

A Series presenting the results of activities sponsored by the NATO Science Committee, which aims at the dissemination of advanced scientific and technological knowledge, with a view to strengthening links between scientific communities.

The Series is published by an international board of publishers in conjunction with the NATO Scientific Affairs Division

A	Life Sciences	Plenum Publishing Corporation
B	Physics	London and New York
C	Mathematical and Physical Sciences	Kluwer Academic Publishers
D	Behavioural and Social Sciences	Dordrecht, Boston and London
E	Applied Sciences	
F	Computer and Systems Sciences	Springer-Verlag
G	Ecological Sciences	Berlin, Heidelberg, New York, London,
H	Cell Biology	Paris and Tokyo

Series C: Mathematical and Physical Sciences - Vol. 256

Late Quaternary Sea-Level Correlation and Applications

Walter S. Newman Memorial Volume

edited by

D. B. Scott
Centre for Marine Geology,
Dalhousie University, Halifax, Canada

P. A. Pirazzoli
INTERGEO - CNRS, Paris, France

and

C. A. Honig
Centre for Marine Geology,
Dalhousie University, Halifax, Canada

Kluwer Academic Publishers

Dordrecht / Boston / London

Published in cooperation with NATO Scientific Affairs Division

Proceedings of the NATO Advanced Study Institute on
Late Quaternary Sea-Level Correlation and Applications
Halifax, Canada
19–30 July 1987

Library of Congress Cataloging in Publication Data

NATO Advanced Study Institute (1987 : Halifax, N.S.)
   Late Quaternary sea-level correlation and applications :
proceedings of the NATO Advanced Study Institute held in Halifax,
Canada, 19-30 July 1987 / edited by D.B. Scott, P.A. Pirazzoli, C.A.
Honig.
      p.   cm. -- (NATO ASI series. Series C, Mathematical and
physical sciences ; vol. 256)
   "Walter S. Newman memorial volume."
   Includes index.

   1. Sea level--Congresses. 2. Geology, Stratigraphic--Quaternary-
-Congresses.   I. Scott, D. B.   II. Pirazzoli, P. A. (Paoio A.)
III. Honig, C. A.   IV. Newman, Walter S., 1927-   . V. Title.
VI. Series: NATO ASI series. Series C, Mathematical and physical
sciences ; no. 256.
GC89.N38 1987
551.7'9--dc19   88-29425
                                                          CIP
ISBN-13: 978-94-010-6880-2   e-ISBN-13: 978-94-009-0873-4
DOI: 10.1007/978-94-009-0873-4

Published by Kluwer Academic Publishers,
P.O. Box 17, 3300 AA Dordrecht, The Netherlands.

Kluwer Academic Publishers incorporates the publishing programmes of
D. Reidel, Martinus Nijhoff, Dr W. Junk, and MTP Press.

Sold and distributed in the U.S.A. and Canada
by Kluwer Academic Publishers,
101 Philip Drive, Norwell, MA 02061, U.S.A.

In all other countries, sold and distributed
by Kluwer Academic Publishers Group,
P.O. Box 322, 3300 AH Dordrecht, The Netherlands.

TABLE OF CONTENTS

PREFACE

A NATO Advanced Study Institute, "Late Quaternary Sea-level Correlation and Applications", was held together with the Final Meeting of IGCP Project 200 in Halifax, Canada, 19-30 July 1987. This Volume is a collection of the NATO Keynote Papers presented at this meeting. The authors of these papers are from seven of the NATO countries - two each from France, the U.K., Canada, and the U.S.A., and one each from Spain, Germany and the Netherlands. With these authors, we are able to assemble work from virtually all of the world's oceans with several different approaches. The Volume is dedicated to Walter S. Newman, one of the best known and best liked sea-level workers of our time who died shortly before this Conference. This Volume contains one of his last contributions and all contributors to this Volume are honoured to be in the company of Walter's last work.

There are several papers from North Atlantic countries dealing with Holocene sea level in a variety of ways. Shennan summarizes data from the U.K. and makes a preliminary effort to place the data in the context of a model. Zazo & Goy present new data from the coast of Spain and place it in a stratigraphical context. Van de Plassche reassesses previous data and adds new data to the very sea-level sensitive Dutch coast. Leatherman uses sea-level information in the Chesapeake Region to assess coastal management problems. Both Fader and Scott et al. present data from the Scotian Shelf of Eastern Canada - Fader uses regional seismic correlations to suggest a 115 m lowering of sea-level; Scott et al. use new data from Sable Island to suggest sea level was no more than 80 m below present at 15 kybp. Giresse presents Holocene and Pleistocene data on the entire African coast and suggests mechanisms to explain differences. Pirazzoli presents worldwide tidal gauge data suggesting that there is no global sea-level trend in the last 50-100 years. Bruckner presents Holocene sea-level data from the Indian coastline, an area seldom reported on in the English literature. Smith presents a paper showing how archaeologists can use sea-level data in the Pacific to speculate movements of early man. Finally, in a paper typical of Walter S. Newman, Newman et al. discuss their huge data bank of C-14 points and how it might be used.

We would like to thank all the financial contributors to this Volume, most notably NATO Scientific Affairs Division who made this combined Symposium and Field Trip and, ultimately this Volume, possible. The Natural Sciences and Engineering Research Council of Canada provided a special conference grant and financial support also came from IGCP Project 200 and the Canadian National Committee for IGCP. Dalhousie University and the Centre for Marine Geology provided accommodations and personnel. Reviewers for the papers are listed at the end of the Volume.

D.B. Scott, C.A. Honig, P.A. Pirazzoli.
June 1988.

WALTER S. NEWMAN (1927-1987) : AN APPRECIATION

      This Volume is very appropriately dedicated to the memory of
one of our most devoted coastal workers and a beloved teacher at
Queens College in the City University of New York. He travelled
widely, often to congresses of the International Union for Quaternary
Research or to field meetings of the INQUA Shorelines Commission, or
in connection with the IGCP sea-level projects (61 and 200), or with
the Friends of the Pleistocene. He had, indeed, dear friends
everywhere: from home in New York City to Japan, the USSR, Israel, or
Australia.

      Walter was a real New Yorker, born and bred. His B.A. was from
Brooklyn College (1950), his M.A. from Syracuse (1959), his Ph.D. from
N.Y.U. (1960). Along the way, he gained experience with the U.S.
Coast Guard & Geodetic Survey in their Lake Mead seismic work; with
the U.S. Army Corps of Engineers in several roles (1951-56) that
included drilling in Bermuda that got him started on sea-level
studies; and with several engineering geological companies. In 1959-
60, he was a research assistant to the late Maurice Ewing at the
Lamont-Doherty Observatory of Columbia University. Subsequently, he
often spent his summers in Quaternary mapping in New England for the
U.S. Geological Survey.

Walter was a natural, born teacher. He loved arguments with heated discussions and was not in the least worried about changing his mind if the evidence tilted that way. He began in 1960 as a lecturer at Queens College, moving up in steps to become chairman and full professor (1978-87). In the last few years he had also become a senior research associate at the Goddard Institute for Space Studies (NASA), where he learned about some of the mysteries of computers.

Jointly, with the writer, Walter had collected a data bank of all the dated samples of coastal significance from the published volumes of Radiocarbon. These he collated on the IBM-PC and learned how to plot them geographically and how to arrange the data in various ways that helped us appreciate some of the problems involved, notably the tide range and neotectonic variables. Many years earlier we had also collaborated in plotting the geodetic relevelling data on a profile from New York City up to the Canadian border and in constructing glacioisostatic recovery maps of the northeastern U.S. As a result we both became convinced about the important role of neotectonics in modulating the picture of mean sea level as perceived from both tide-gauge data and Holocene stratigraphy. His computer-modelling of postulated geoid deformation will stand as the first pioneer experimentsin this field that were based on real (as opposed to hypothetical) evidence: "paleogeodesy" is now among his well-deserved claims to fame.

One of Walter's last projects was to help organize the "birthday symposium" and edit the resultant book, "Climate: History, Peridocity, and Predictability" (edited by Rampino, Sanders, Newman and Konigsson, 1987). Through his life he was wonderfully supported by his wife, Marian, and was understandably proud of his two fine daughters. We all miss him very much.

Rhodes W. Fairbridge
July 1988.

HOLOCENE SEA-LEVEL CHANGES AND CRUSTAL MOVEMENTS IN THE NORTH SEA REGION : AN EXPERIMENT WITH REGIONAL EUSTASY

Ian Shennan
Department of Geography
University of Durham
Durham DH1 3LE
United Kingdom

ABSTRACT. Eighteen relative sea-level curves from the North Sea region are used to test the concept of regional eustasy. Net uplift/subsidence curves for each location are derived but no single baseline, regional eustatic, curve will account for all the residuals identified. The simplest conclusion is that dynamic sea-level factors have not remained constant on the scale of the North Sea during the Holocene.

1. INTRODUCTION

The coastal areas bordering the North Sea have been systematically studied for over 100 years to identify the factors of local and regional significance controlling shoreline processes during the Holocene. The sea-level indicators range from uplifted shoreline sequences and lake basin sediments in Scandinavia and Scotland to thick intercalated sequences of marine, brackish, freshwater and semi-terrestrial sediments in the estuaries bordering the southern North Sea. The collection of empirical data has been undertaken by a wide range of individuals and groups from numerous universities, state and federal survey and research institutions, providing the potential of a large database, perhaps unequalled for any other area of the world of comparable size. Synthesis of the data at the regional scale has hitherto been largely precluded by the volume and quality of the data. This paper attempts to remidy the situation and adopts a regional approach. It is a modified update of a longer contribution published elsewhere (Shennan 1987a).
    Extensive data gathering is typical of a scientific discipline during its early stages of development with explanation via model building and hypothesis testing coming later. Many sea-level studies in Europe have remained

1

D. B. Scott et al. (eds.), Late Quaternary Sea-Level Correlation and Applications, 1–25.
© *1989 by Kluwer Academic Publishers.*

essentially inductive in their approach with little attempt
to put together knowledge of the separately known events,
ultimately to lead to theory construction and thence to
explanation (Tooley 1987). However the increase,
particularly during the period of the two IGCP sea-level
projects, 61 and 200, in the number of reliable radiocarbon
dates on Holocene marine episodes, together with more
standardised methods of data collection, analysis and
classification enable alternative routes to explanation in
sea-level studies to be attempted (Shennan 1983).

The aim in this paper is not to provide a summary of
research papers but to investigate some of the methods that
can be applied in synthesising the available data. Similar
methods have been discussed elsewhere (Shennan 1987a). To
summarise published data, judgements must be made using a
database poorer than that available to the original research
worker, for example full stratigraphic details, sampling
problems and laboratory analyses are rarely accessible.
Even if the required details were published it would be very
difficult to synthesise all the data on a regional scale
given the resources currently available to individual
research workers, either in isolation or within the sphere
of cooperation of international projects such as IGCP 61 and
200. The explosion of detailed data as a result of
interdisciplinary research, whilst making the data more
reliable, particularly in helping to describe local scale
changes, makes the task of synthesis increasingly difficult.
The problem to be addressed is how best to utilise data
which hass been collected at one temporal and spatial scale,
e.g. for local archaeological investigations or geological
surveys, to answer questions related to processes effective
at a different resolution or regional scale.

Mörner (e.g. 1976a, 1987) has introduced the concept of
regional eustasy following observations and arguments about
the temporal and spatial variability of the oceanic geoid.
A new definition of the term eustasy has been proposed
(Mörner 1987 p.338), which now covers tectono-eustasy,
glacial-eustasy, geoidal eustasy and dynamic sea-level
changes. Each of these will be of greater or lesser
significance over different temporal and spatial scales.
Peltier (e.g. 1982, 1985) employs a different approach,
using a spherical earth model reconstructions of ice sheet
retreat to provide numerical predictions of relative
sea-level change for any location. This approach is not
considered in detail in this paper since work is only now
underway to compare the numerical predictions for sites in
the United Kingdom with the empirical relative sea-level
data.

Mörner (1980) has suggested that the North Sea region
can be viewed as an immense sea level laboratory in which
various approaches to isolating the different eustatic and

other variables can be experimented with. In this paper
one of these approaches is taken whereby the proposed
regional eustatic sea-level curve is tested as a model to
explain local relative sea-level changes and reveal the
regional pattern of crustal movements around the North Sea.
In order to identify the limits of resolution that may be
obtained in fitting local records to the regional eustatic
model, data from three sources are analysed to varying
degrees. Firstly sea-level index points for the Fenland
area of eastern England, derived from empirical data
collection in the field and subsequent laboratory analyses;
secondly radiocarbon dated sea-level index points collected
from other areas of the U.K. supplied by members of the U.K.
Working Group of ICGP Project 200; and finally published
sea-level curves from the coasts of the North Sea (figure
1). Collectively these three analyses can be used to
illustrate how databases of different complexity can be
integrated.

It should be stressed that the approach taken in this
paper is one experiment that can be performed with the
currently available data, the strengths and weaknesses will
be discussed later in this context.

2. THE EUSTASY MODEL

The initial assumption is that the regional eustatic curve
proposed by Mörner (1971, 1976b, 1980, 1984, see also
Shennan 1987a) can be used as a benchmark for comparison of
sea-level records from different locations around the North
Sea. To facilitate experimentation, using different
numerical values in various analyses the regional eustatic
curve was stored on computer file as a sequence of altitudes
between 8800 and 1000 BP using the fixed points for the
maxima and minima of each sea-level oscillation (Shennan
1987a). The remaining points were interpolated using a
half wavelength sine curve passing through each pair of
alternating maxima and minima. The error induced by this
approximation is mostly of the order of a few centimetres,
and not significant in comparison with the other factors
considered given the specified temporal and spatial scales.
A coarser approximation, based on the maxima only, has also
been used (Shennan 1987a) to evaluate the stability of the
computed relationships dependent upon changes in the
original assumptions.

3. REGIONAL EUSTASY AND LOCAL FACTORS - EASTERN ENGLAND
CASE STUDY

Relative changes in sea level and the nature of

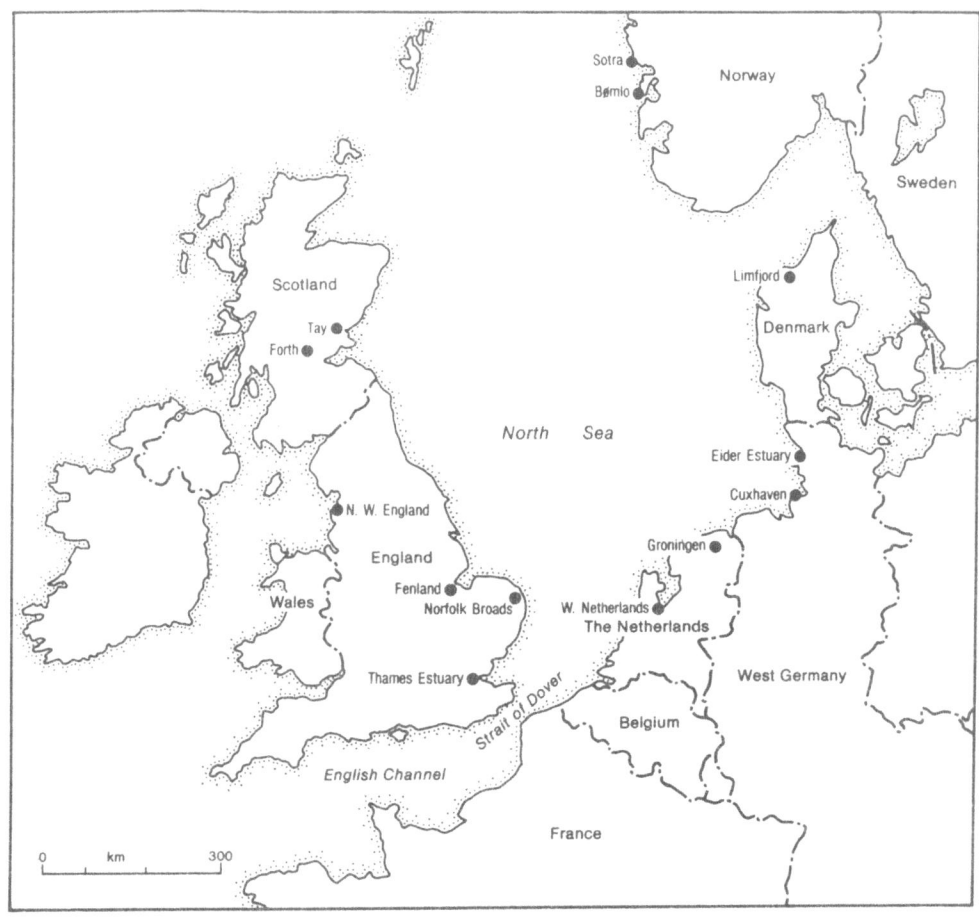

1. A map of the North Sea and adjacent countries, showing
    the locations of sites discussed in the text.

environmental changes in the Fenland embayment of eastern
England over the last 6500 radiocarbon years have been
studied in some detail (Shennan 1986a,b). The relative
sea-level band derived from the field data is shown in
figure 2. This band indicates short term interruptions in
the general sea-level rise and these are assumed to
represent important, though low amplitude events. The
variations of the sea-level band have been compared to
lithostratigraphic, biostratigraphic, chronostratigraphic
and archaeological data and alternating periods of dominant
positive and negative sea level tendencies have been
identified (Shennan 1986b). These periods are shown WI-WVI
and FeI-FeVI respectively on figure 2.

Using the proposed regional eustatic curve as the basis
for calculation the average linear rate of subsidence for
the last 6500 radiocarbon years is estimated at c.0.91m/1000
years (Shennan 1986b). The scatterplot of data points
showing subsidence against time indicated that a linear rate
was an adequate summary of the data currently available
(Shennan 1986b). There were insufficient data to constrain
any calculations of a curvelinear rate. An increase in the
size of the dataset available (Shennan in prep.) may
indicate a non-linear rate is justified.

Different environmental systems will react in
dissimilar ways to changes in rates of controlling processes
or inputs and outputs for energy or materials. Therefore
low amplitude changes in sea level will not be recorded in
all types of environment. Mörner (e.g. 1980, 1984) records
9 maxima and 9 minima in the regional eustatic curve between
6500 and 2000 BP, compared to only 4 periods of reduced or
negative rates of rise shown in the sea-level band for the
Fenland. This would suggest that the Fenland coastal
sedimentary system, showing alternating marine tidal flat
deposits and peat beds, does not react equally to all
eustatic oscillations. By taking the linear solution for
crustal subsidence, 0.91m/1000 years, and the regional
eustatic curve to calculate a local relative sea-level curve
the predicted and observed responses recorded in the Fenland
can be analysed. The predicted relative sea-level curve is
shown in figure 3, and can be compared to the sea level
band, rescaled and redrawn, derived from the empirical field
data (figure 2).

The predicted and observed data mainly show very close
agreement. It has been suggested that a time resolution for
correlations no finer than 100-200 years should be allowed
for in between-area analyses (Mörner 1980a, Shennan 1987a)
and a similar error has been specified for the sea-level
chronology from the Fenland (Shennan 1986b). Nevertheless
small shifts of this magnitude in the timing of the maxima
and minima on either the regional eustatic curve or local
sea-level band would remove any disagreement. The effect of

such small changes in the parameters has been illustrated elsewhere (Shennan 1987a).

The alternations in dominant sea-level tendencies (figures 2 and 3) clearly have an eustatic origin. Each period of negative tendencies of sea-level movement in the Fenland area comes at the end of a prolonged period of reduced rise or a period of small oscillations in the regional eustatic curve modified for the local subsidence rate. Each period of negative tendencies is halted by a subsequent major rise of sea level. Short periods of rapid change seem important in delaying the registration of a negative tendency. Rather than responding immediately to changes in sea-level movement the Fenland environmental system requires some time to adapt to the new parameters. Other estuaries or other sedimentary environments need not show a similar lag-time, further complicating the process of between area correlation (c.f. Shennan 1987b), although major consistencies will still be identified (Shennan et al 1983).

This comparison of predicted and observed sea-level changes and tendency chronology indicates the clear regional eustatic origin of the local changes. These data can then be used to compute changes in the rate of predicted regional eustatic sea level modified by the local subsidence factor (figure 3) and the net shoreline response (table I). Future changes in the rate of sea-level rise (table II) predicted in various scenarios (Barth and Titus 1984) are in excess of the thresholds for the initiation of significant shoreline retreat in the past, as shown by comparison of tables I and II (discussed further in Shennan 1987c).

Table I : Maximum rates of sea-level change (estimated 50yr mean, mm/yr) associated with the end of periods of regressive overlap in the Fenland, U.K.

Age (BP)	Eustasy model 1	Eustasy model 2
6200	14.5	14.5
5400	10.0	9.6
4200	8.0	7.5
3000	5.2	2.9
2500	8.3	6.3

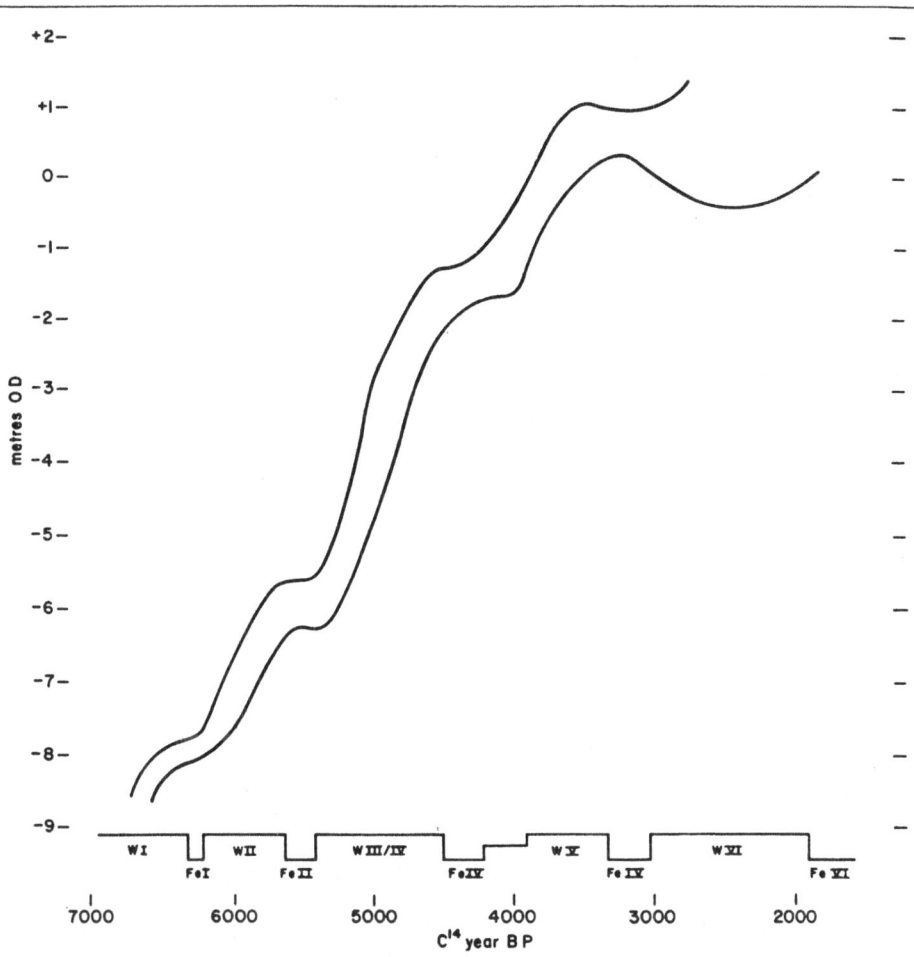

2. Relative sea-level band (mean high water of spring
tides) for the Fenland, U.K., derived from an analysis
of field and laboratory data (Shennan 1986b). Current
MHWST is 3.32 ± 0.13m above mean tide level in the Wash
estuary, and 3.80 ± 0.07m above Ordnance Datum.

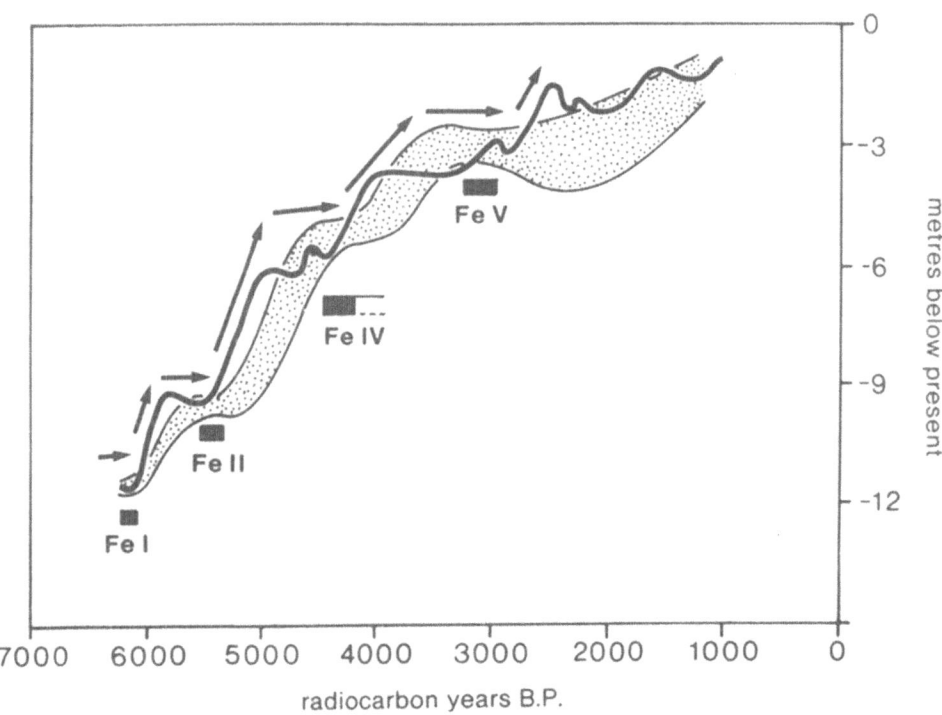

3. Comparison of the computed sea-level curve, thick line, and the empirically derived sea-level band from figure 2. Horizontal black bars indicate periods FeI-FeV and the arrows emphasise the correlation between each period of negative tendency of sea-level movement, FeI-FeV, and the major trends of predicted sea level.

Table II : Rates of sea-level rise (50 year mean, mm/yr) for various scenarios 1975-2100 AD, adapted from Barth and Titus (1984).

Scenario	1975 - 2025	2000 - 2050	2050 - 2100
conservative	2.6	3.8	4.6
mid-range low	5.2	8.2	18.4
mid-range high	7.8	12.8	27.6
high	11.0	20.0	45.6

This analysis has shown that the regional eustatic curve is a useful predictive model for evaluating the local record from the Fenland. There are discrepancies, but these are deemed to be within the imprecision of the raw data, i.e. altitudinal reconstruction limits and radiocarbon dating errors. However, if these discrepancies, or residuals, occur consistently when similar techniques are applied elsewhere then non-random factors must be considered. The next stage of analysis is to use the regional eustatic model at the spatial scale of the North Sea.

4. REGIONAL EUSTASY AND THE NORTH SEA AREA

A regional approach to analysing the interaction between the various eustatic factors, including dynamic sea-level changes (Mörner 1987), and crustal factors requires the comparison of data collected by various authors. The main sources of data used here are published sea-level curves, supplemented in the examples for the UK by data collected for the IGCP-200 radiocarbon databank. One unpublished curve from the Norfolk Broads (Alderton 1983) has been adopted, using IGCP-200 databank entries.

A reasonable spatial cover of the North Sea region, and adjoining areas, can be gained from 14 relative sea-level curves (figure 4). Relative sea-level altitudes and values for net uplift or subsidence for each curve at 500 radiocarbon-year intervals were obtained using the regional eustatic curve as a baseline (further details in Shennan 1987a). With a regional eustatic curve showing fluctuations and a relative sea-level curve also containing fluctuations the combination of the two curves may be affected by the extent to which the fixed 500-year datapoints coincide with the maxima or minima on either curve. To assess the effect the analyses were duplicated

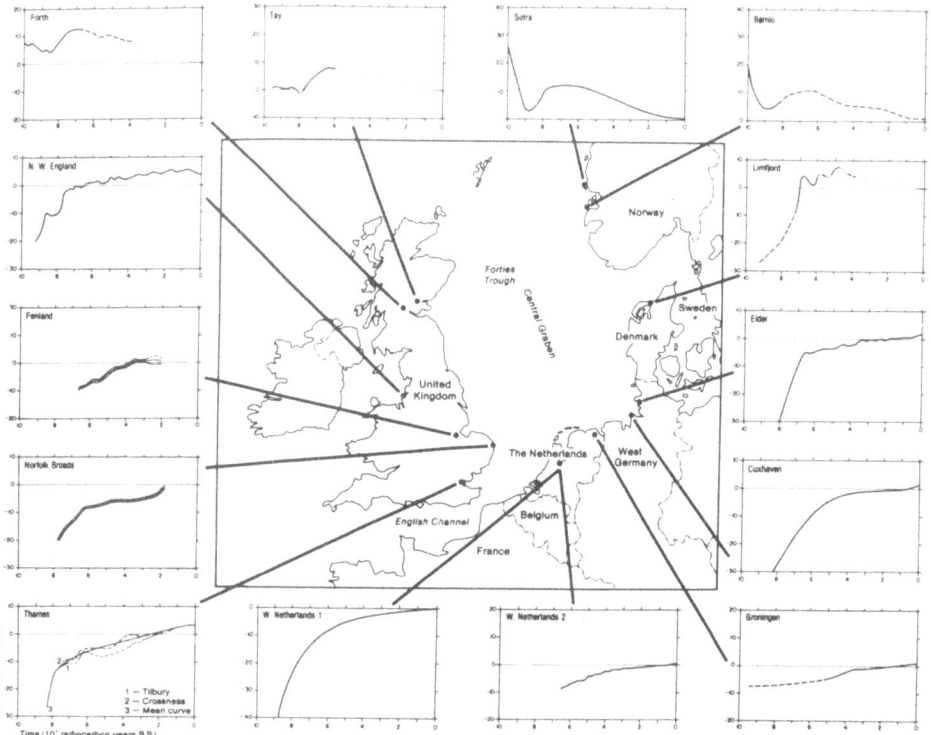

4.   Relative sea-level curves from around the North Sea,
       and one from North-West England, drawn on a uniform
       scale (from Shennan 1987a).   Dashed lines indicate a
       decrease in the reliability of the curve noted by the
       original author (reference in the text).   The dotted
       section for the Forth curve indicates even lower
       reliability noted by the author;   the dotted section
       for the Tay and Limfjord are extrapolations made for
       this analysis.

using a modified regional eustatic curve, interpolated by a straight line between each successive pair of maxima, that is the short term fluctuations have been removed.

An alternative method to using published curves, which clearly include interpolations made by the original authors, is to analyse the individual sea-level index points (e.g. Flemming 1982, Newman and Baeteman 1987). It has not been possible within the two IGCP Projects to accumulate a sufficiently reliable and international database to develop this route of analysis to date.

Scatterplots showing apparent uplift and subsidence were produced for each location. Each plot was inspected and the time period identified where a linear trend was apparent. The linear trend was calculated in eight separate ways, representing the possible combinations for the following three conditions: i) the linear regression calculated with a constant term or forced through the origin, ii) the eustatic curve with fluctuations or interpolated as a straight line between the maxima only, iii) a radiocarbon or sidereal timescale. Further details of the combinations are discussed by Shennan (1987a) and only those analyses based on the radiocarbon timescale are considered here, i.e. four methods of calculating the apparent linear trend.

The results of the linear regression analyses are shown in table III. The constant term shows how close the trend passes to the origin. A large constant term may indicate a number of factors, e.g. the statistical effect of calculating a linear function for a non-linear relationship, or a misinterpretation regarding the relationship of the sea-level index points or the sea-level curve to the correct tidal level. Given the accuracy of the data and their abstraction a constant term less than $\pm$ 1.0m is probably not important.

The results for each area are now outlined, with the curves used for data collection referenced.

4.1. The Fenland, United Kingdom (Shennan 1986b)

Relative sea-level changes in the Fenland are presented in the source publication by a sea-level band and three analyses have been carried out, using the maximum, minimum and mean values at each time point. The results for the 'minimum' analyses show the weakest correlation coefficients and largest constant terms (table III), indicating that the minimum line of the sea-level band is unlikely to relate to a constant reference water level, and is certainly not equivalent to mean high water of spring tides (the reference water level on the original curve). The Fenland data were not adjusted for the effects of sediment consolidation and

Table III : Summary of linear trends of uplift and subsidence in the North Sea Region. Note : cases used refer to those data identified as indicating a linear trend on the scatterplots. The four results for each area are based on the following methods of calculating a linear trend; (i) eustatic curve with regressions, linear regression with constant term; (ii) eustatic curve with regressions, linear regression through origin; (iii) eustatic curve with no regressions, linear regression with constant term; (iv) eustatic curve with no regressions, linear regression through origin; 'r' = the regression coefficient. The values used in the calculations were obtained from the original references (see text), the scatterplots are diagrammatic summaries of the results and do not indicate the resolution of the original data.

	cases used (yr BP)	constant (m)	rate (m/1000yr)	r
Tay Estuary	0-5000	-0.35	1.60	0.98
			1.46	
		-0.41	1.53	0.98
			1.37	
Forth Estuary	0-5000	-1.23	2.42	0.97
			1.92	
		-1.29	2.34	0.96
			1.83	
North-west England	0-5500	0.78	-0.00	-0.06
			0.29	
		0.69	-0.09	0.25
			0.16	
Fenland : maximum	0, 2000 - 5500	0.14	-0.87	-0.88
			-0.84	
		-0.05	-0.94	-0.91
			-0.95	
Fenland : mean	0, 2000 - 5500	-0.41	-0.86	-0.87
			-0.97	
		-0.61	-0.93	-0.89
			-1.08	
Fenland : minimum	0, 2000 - 5500	-0.96	-0.85	-0.83
			-1.10	
		-1.16	-0.92	-0.85
			-1.22	

Table III continued

	cases used (yr BP)	constant (m)	rate (m/1000yr)	r
Norfolk Broads : maximum	0, 1500, 2000, 6000 – 7500	-0.19 -0.47	-0.47 -0.51 -0.46 -0.57	-0.97 -0.93
Norfolk Broads : mean	0, 1500, 2000, 6000 – 7500	-0.40 -0.68	-0.52 -0.62 -0.52 -0.68	-0.94 -0.90
Norfolk Broads : Minimum	0, 1500, 2000, 6000 – 7500	-0.83 -1.11	-0.55 -0.74 -0.55 -0.80	-0.88 -0.84
Thames Estuary : mean	0, 1000 – 5000	-0.01 -0.11	-1.12 -1.13 -1.19 -1.23	-0.96 -0.98
Thames Estuary : Tilbury	0, 1000 – 5000	0.83 0.73	-2.00 -1.69 -2.06 -1.79	-0.97 -0.98
Thames Estuary : Crossness	0, 1000 – 5000	-0.05 -0.15	-1.16 -1.18 -1.22 -1.28	-0.90 -0.93
Western Netherlands 1	0 – 7000	0.27 0.15	-0.49 -0.42 -0.55 -0.51	-0.88 -0.95

Table III continued

	cases used (yr BP)	constant (m)	rate (m/1000yr)	r
Western Netherlands 2	0 - 5000	0.13	-0.53 -0.47	-0.87
		0.75	-0.60 -0.57	-0.96
Groningen	0 - 5500	-0.16	-0.55 -0.61	-0.83
		-0.08	-0.71 -0.74	-0.98
Cuxhaven	0 - 6000	-0.32	-0.65 -0.76	-0.86
		-0.27	-0.79 -0.86	-0.94
Eider Estuary	0 - 5000	-0.43	-0.61 -0.78	-0.76
		-0.49	-0.68 -0.88	-0.89
Limfjord	0 - 5000	-0.64	1.69 1.43	0.96
		-0.71	1.62 1.34	0.45
Bomlo	0 - 5000	-0.45	1.56 1.38	0.98
		-0.51	1.49 1.29	0.98
Sotra	0 - 5000	-1.57	2.29 1.66	0.96
		-1.63	2.21 1.57	0.96

this is likely to have affected the minimum line of the
sea-level band more than the mean or maximum estimates.
        The rates shown on the figures for each area (figure 5)
are the average of the four linear regressions given in
table III.
        For the Fenland the average linear rates of subsidence
calculated for the past 5500 radiocarbon years are 0.90 ±
0.05m/1000 year for the maximum line and 0.96 ± 0.09m/1000
year for the mean values. These rates compare to
0.91m/1000 year, calculated from a different method of
analysis (Shennan 1986b), for the past 6500 years. There
are no consistent deviations from the linear trends (figures
5) except for some indication of a reduced rate from
6500-5500 BP (also noted in Shennan 1986b). A similar
feature is noted in some of the other curves discussed
below.

4.2. The Norfolk Broads, United Kingdom (adapted from
Alderton 1983, and IGCP-200 databank)

As with the Fenland the 'minimum' analysis can be rejected.
For both the 'maximum' and 'mean' scatterplots (figure 5) no
linear solution can explain all the data points. Two
factors additional to linear subsidence appear important.
Firstly the original index points are from sites up to 30km
from the present coast, along narrow river valleys, with a
spit diverting the river south as it reaches the present
coast. Changes in tidal range, presently 1.9m for spring
tides at the open coast, are likely to have occurred in the
past, affecting each sampling site to varying extent. An
increase in range would have been shown as a positive
residual from a linear trend. Secondly for the period
after 5000 BP the data points are mostly from intercalated
peats overlying thick sequences of unconsolidated silts,
clays and peats, which are more susceptible to consolidation
effects than the earlier index points from basal peat
transgressive overlaps.
        The combination of tidal range variation, giving
positive residuals 7500-5000 BP, consolidation, negative
residuals 5000-1500 BP, and a linear rate of subsidence
similar to the Fenland is indicated in figure 5 as a dashed
line. By ignoring any tidal effect and giving greater
importance to consolidation (dotted line) average rates of
subsidence are estimated as 0.59 ± 0.08 m/1000 year (mean).
The consolidated effect is not so important in the Fenland
where index points from near the base of the unconsolidated
sequence are used at intervals throughout the time period
considered, whereas for the Norfolk Broads such index points
are only available in the early part of the sequence.

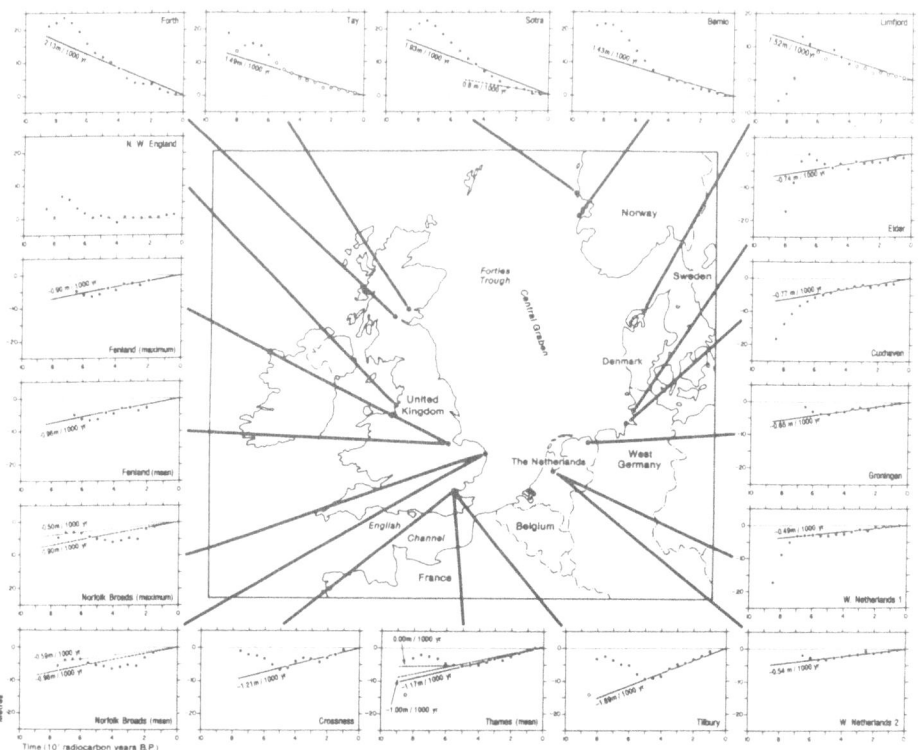

5. Computed uplift and subsidence from sites around the North Sea (from Shennan 1987a). Open circles indicate that the data were obtained by the author by extrapolation. The rate of uplift/subsidence is the mean value obtained by four methods of calculation (table III). The data points on the scatterplots are for the eustatic curve with no regressions and therefore they will tend to appear slightly below the mean rate line shown, which is calculated from the data using the eustatic curve with regressions as well. These data were excluded from the scatterplots for clarity. The lines on the scatterplots have been extended in some cases for ease of labelling but the linear rates were calculated for the periods given in table III. Non-linear relationships are evident for many areas as noted in the text.

4.3. The Thames Estuary, United Kingdom (Devoy 1979)

Devoy (1979) discusses the clear within-estuary variations, caused only in part by sediment consolidation, tidal amplitude and river discharge, and attributes the difference between the Crossness and Tilbury curves, sites 16km apart, to differential crustal subsidence. Devoy also draws a mean curve for the estuary.

Comparisons of the subsidence scatterplots (figure 5) reveal the magnitude of the differential crustal subsidence. Linear trends are not apparent for the whole of the time period but only for the last 5000 years: subsidence of 1.89 ± 0.16m/1000 yr at Tilbury, 1.2 ± 0.05m/1000 yr at Crossness and 1.17 ± 0.05m/1000 yr for the mean curve.

Prior to 5000 BP there are significant trends away from these linear solutions. The field data are considered to be of highest quality (Devoy 1979, Shennan 1987a) thus various possible explanations must be hypothesised. The stratigraphic and spatial distribution of the data points reveals that consolidation will not have had a significant effect. Devoy (1982) estimates between 0.3 to 0.9m for some of the index points whereas the deviations from the linear trend for the index points older than 5000 BP are much greater. Tidal parameters are unlikely to have remained stable throughout this period, during which the English Channel - North Sea connection was completed and the present circulation pattern in the region established but whether such factors would explain the 7-11m residual at 7500 BP is arguable, even with a reduction of 1m by allowance for consolidation. Some tidal effect may be apparent, otherwise local uplift 8000-5000 BP is signified (figure 5) and this is difficult to explain. A tidal effect combined with zero subsidence prior to 5500 BP followed by linear subsidence between a minimum of 1.0 to 0.03m/1000yr (allowing for consolidation effects) to 1.17 ± 0.05m/1000yr are indicated on figure 5.

4.4. North-West England, United Kingdom (Tooley, 1978)

No complete sea-level curve is yet available from the North-East coast of England so this curve is included to illustrate the crustal movements between the subsiding areas to the south and the uplifted areas further north. Figure 5 shows essentially a zero rate of uplift for the past 5000 or 6000 yr with an exponential decrease prior to then. The oldest two index points appear as significant residuals from any hypothesised exponential curve for glacio-isostatic uplift and these residuals can only be partly explained by considering error estimates in the original data (Tooley 1982, Shennan 1987a).

4.5. Forth Estuary, United Kingdom (adapted from Sissons 1967 and Sissons and Brooks 1971)

The original curve is poorly established for the last 4000 yr and a linear trend (figure 5) is not a suitable summary for most of the data. The average of the rates given in table III (note the large constant terms) is 2.13 ± 0.30m/1000 yr and is for general guidance only. It should be noted that, as with the North-West England, the two oldest data points appear a few metres too low from an uplift curve showing a curvelinear decline.

4.6. The Tay Estuary, United Kingdom (Cullingford et al 1980)

The original relative sea-level curve is too incomplete to be used for anything other than an estimate of the broad trends of uplift (table III and figure 5). The data point for 8000 BP was interpolated from the original relative sea-level curve, nevertheless deviations from a simple curvelinear decline, similar in form to the previous two areas, can be noted.

4.7. Sotra, Norway (Kaland et al 1984)

The linear solution is a poor summary of these data, whether expressed for the last 5000 or last 2000 yr (table III and figure 5). The oldest data point is some metres too low: the next oldest is too low on the diagram shown (figure 5) but if the data points from the regional eustatic points with regressions had been reproduced it would have appeared on the line of the curve.

4.8. Bomlo, Norway (Kaland 1984)

The curvelinear relationship for the 8500 yr period is clear on figure although a linear summary, 1.43 ± 0.11m/1000 yr of uplift, appears adequate for the last 5000 years. This is possible corroboration of the 'double factor' uplift of Scandinavia (Mörner 1980b), but may also be a limitation of the data used here. Once more the oldest index point appears a few metres too low.

4.9. Limfjord, Denmark (Peterson 1984)

Data limitations (see Shennan 1987a) mean that the indicated linear rates of uplift, 1.52 ± 0.16m/1000 yr (table III and figure 5) are only a general guide for the period 0-5000 BP, and for 7500-8500 BP the data points appear anomalous, although the altitudinal accuracy of this early part of the original curve was poor.

4.10. The Eider Estuary, West Germany (Behre et al 1979)

The small fluctuations in the original curve are inferred
from botanic and stratigraphic evidence. The linear trend
for 0-5000 BP of 0.74 ± 0.11m/1000 BP subsidence is a good
summary of the data, the variability of which is exaggerated
due to the accuracy with which the data could be obtained
from the published curve, i.e. 0.40m. Prior to 7500 BP the
curve is obtained from samples offshore, from areas of
possible enhanced subsidence. As with other areas
discussed the 6500-5000 BP period reveals a reversed slope
relationship. Due to the resolution of the data for this
area it is unclear whether this reflects a tidal, crustal or
consolidation effect.

4.11. Cuxhaven, West Germany (Linke 1982)

Unlike Behre et al (1979) Linke produced a smooth mean tide
level curve, arguing that apparent fluctuations were due to
variation between very stormy phases and storm-free phases.
Nevertheless the subsidence scatterplots (figure 5) shown a
linear trend for 0-6000 BP of 0.77 ± 0.08m/1000 yr,
indistinguishable from the Eider rate, with greater
subsidence prior to then. As with the Eider and Western
Netherlands analyses the early part of the curve is based on
data from the offshore zone, and revealing quite different
subsidence effects.

4.12. Groningen District, The Netherlands (Roeleveld 1976)

The relative sea-level curve is a summary of many
stratigraphic and radiocarbon analyses but drawn through
only ten fixed points. A linear subsidence rate of 0.65 ±
0.08m/1000 yr for 0-5500 BP is evident (table III, figure
5). For 6500-5500 BP this trend is reversed, a feature
noted for some other areas also. It is possible that in
this instance the method for curve construction is
responsible in part (see Shennan 1987a), but tidal effects
and variation in the rate of subsidence may be represented.

4.13. The Western Netherlands (two curves, 1: Jelgersma
1980, corrected by van de Plassche 1980; 2: Louwe Kooijmans
1974)

These two curves are composites of data obtained from a
relatively large geographical area. For the period 0-6500
BP the two curves reveal very similar subsidence
characteristics: 0.49 ± 0.05m/1000 yr for curve 1, 0-7000
BP; 0.54 ± 0.05m/1000 yr for curve 2, 0-5000 BP. The
apparent reduced subsidence for the early part of curve 2
may be the result of using one, rather imprecise, data point

at the extreme of the data range, although since similar
features are revealed for other areas alternative
explanations should not be rejected.

The part of curve 1 pre-7000 BP is obtained from
samples up to 160km offshore, therefore probably affected by
different subsidence histories and not requiring further
comment here.

4.14. Discussion

This regional-scale approach shows how relative sea-level
curves can be used to illustrate the possible interaction of
crustal movement, tidal variations and local sediment
consolidation. The isolation of all of these variables are
dependent on the assumption that a single sea-level curve
can be adopted as a base line and that the curve proposed by
Mörner is appropriate. The problem of scale regarding
Mörner's new definition of eustasy (Mörner 1987) is now
relevant. Whilst the North Sea may be hypothesised a
unified area taking into account the combined effects of
tectono-eustasy, glacial eustasy and geoidal eustasy
(hypothesis 1), that dynamic sea-level changes would have
acted uniformly over the area for the last 8000 years
(hypothesis 2) would appear less likely. This must be
considered further in the future using the numerical
approach of Peltier. Meanwhile, using the present approach,
if hypotheses 1 and 2 are both correct, but the proposed
regional eustatic curve contains an error, then anomalies
will be recorded in the analysis of each area (sections 4.1
- 4.13) for the same time period.

Most of the subsidence plots from the southern North
Sea (sections 4.1, 4.2, 4.3, 4.10, 4.12, 4.13) revealed a
decreased rate of subsidence or even apparent uplift for the
period c. 6500-5000 BP (figure 5). These residuals
remained during all analyses, i.e. eustatic curve with or
without oscillations, radiocarbon or sidereal timescale (see
Shennan 1987a). If the regional eustatic curve is in error
any proposed correction would have produced increased
positive residuals in the uplift curves from Denmark,
Norway, Nort-West England and Scotland. Alternative
explanations would include the imprecision of the relative
sea-level curves, possible for only some of both the
subsiding zone and uplifted zone curves, sediment
consolidation, probable in at least one example; changes in
the dynamic sea-level factors, certainly a possible factor
as the bathymetry of the coastal zone and estuaries of the
North Sea developed; or a variation in the rate of
subsidence, perhaps due to the increasing hydro-isostatic
load. The large residuals for the Thames estuary may
indicate that all these factors operated in this area, and
with different effects over relatively short distances.

The uplift plots (figure 5) reveal that the data points for 8500 BP show consistently low uplift values in comparison to an expected curvelinear decrease to the present day. This factor had also been noted by Mörner (1976b) in his analysis of the North West England data. Precision of the data may be lower for this time period (see Shennan 1987a) and any slight alteration, i.e. the interpretation of the radiocarbon dates to include their error terms for either the eustatic curve or the relative curve may be sufficient to remove any significant residual from a smooth uplift curve, when converted to a sidereal timescale. More detailed investigations, using new data and new methods, will be required to confirm or refute this.

Given the current precision and range of data hypothesis 1 cannot be reliably rejected. The deviations noted for 6500-5000 BP and 8500-8000 BP may be evidence to suggest that either the regional eustatic curve is in error, (but this does not explain the fact that no single baseline curve will adequately account for all the residuals shown in the present analysis) or that the geoid has changed configuration, on the scale of the North Sea during the Holocene. The simplest explanation is to reject hypothesis 2, that the dynamic sea-level factors have not remained constant on the scale of the North Sea during the Holocene.

Therefore the regional eustatic model, taking into account the new definition proposed by Mörner (1987a), must be deemed in error in terms of its spatial application. Whether such a broad definition for eustasy will be useful awaits to be seen (see reference to ongoing work, section 1.0) but clearly the definition of relevant temporal and spatial scales is essential in discussing the different factors. Users of such terms will have to be more explicit than in the past since eustasy now appears to include changes operating on all scales from global to estuarine.

5. CONCLUSIONS

One approach to the regional analysis of relative sea-level data has been described here using mainly published sea level curves. This approach can be significantly enhanced by using both individual sea-level index points and digitised sea-level curves. This can only progress with an integrated computer-based relational database. The flexibility and repeatability of such methods have been evident in carrying out recent research (e.g.Shennan 1983, 1987a, b) even though neither all the dead ends or all the exciting openings have been alluded to here. They offer very many possible lines for future research but the effort required to accumulate the required data into computer compatible form appears to be beyond the capabilities of the cooperation fostered within the IGCP sea-level projects. The U.K. Working Group of IGCP 200 undertook as one of their objectives to complete the database of sea-level index points for the U.K. Walter Newman was the only other participant, along with his co-workers, interested in the active transfer and use of the computerised database (see van de Plassche 1987 p.60). His imaginative and investigative research (e.g. Newman et al 1980, 1981. Newman and Baeteman 1987) will be missed greatly. Without renewed and increased effort and cooperation regional sea-level analyses will be held back by a lack of suitable, comprehensive, and readily available data.

REFERENCES

Alderton, A. 1983: Flandrian vegetational history and sea-level change of the Waveney Valley. Unpublished Ph.D. Thesis, University of Cambridge.

Barth, M.C. and Titus, J.G. (eds) 1984: Greenhouse Effect and Sea-level Rise: a challenge for this generation. New York: Van Nostrand Reinhold

Behre, K.-E., Menke, B. and Streif, H. 1979: 'The Quaternary geological development of the German part of the North Sea.' In E. Oele, R.T.E. Schüttenhelm and A.J. Wiggers (eds), The Quaternary History of the North Sea. Acta Univ. Ups. Symp. Univ. Ups. Annum Quingentesimum Celebrantis, Uppsala, 2, 85-113.

Cullingford, R.A., Caseldine, C.J. and Gotts, P.E. 1980: 'Early Flandrian land and sea-level changes in Lower Strathearn.' Nature, 284, 159-61.

Devoy, R.J.N. 1979: 'Flandrian sea-level changes and vegetational history of the lower Thames Estuary.' Phil. Trans. R. Soc. Lond., B285, 355-407.

Devoy, R.J.N. 1982: 'Analysis of the geological evidence for Holocene sea-level movements in south-east England.' Proc. Geol. Ass., 93, 65-90.

Flemming, N.C. 1982: 'Multiple regression analysis of earth movements and eustatic sea-level changes in the United Kingdom in the past 9000 years.' Proc. Geol, Ass., 93, 113-25.

Jelgersma, S. 1980: 'Late Cenozoic sea-level changes in the

Netherlands and the adjacent North Sea Basin.' In N.-A. Mörner (ed), Earth Rheology, Isostasy and Eustasy, Chichester:John Wiley, 435-47.

Kaland, P.E. 1984: 'Holocene shore displacement and shorelines in Hordaland, western Norway.' Boreas, 13, 243-58.

Linke, G. 1982: 'Der Ablauf der holozänen Transgression der Nordsee aufgrund von Ergebnissen aus dem Gebiet Neuwerk/Scharhorn.' Probleme d. Küstenforsch i. südl. Nordsegebiet, 14. 123-57.

Louwe Kooijmans, L.P., 1984: The Rhine/Meuse delta: Four studies on its prehistoric occupation and Holocene geology. Leiden: E.J. Brill.

Mörner, N.-A. 1971: 'Eustatic changes during the last 20 000 years and a method for separating the isostatic and eustatic factors on an uplifted area.' Palaeogeogr., Palaeoclimatol. Palaeoecol. 9, 153-81.

Mörner, N.-A. 1976a: 'Eustasy and Geoid Changes.' J. Geol., 84, 123-51.

Mörner, N.-A. 1976b: 'Eustatic changes during the last 8000 years in view of radiocarbon calibration and new information from the Kattegatt region and other north-western European coastal areas.' Palaeogeogr., Palaeoclimatol., Palaeoecol., 19, 63-85.

Mörner, N.-A. 1980a: 'The north-west European 'sea-level laboratory' and regional Holocene eustasy.' Palaeogeogr., Palaeoclimatol., Palaeoecol., 29, 281-300.

Mörner, N.-A. 1980b: 'Eustasy and geoid changes as a function of core/mantle changes.' In N.-A. Mörner (ed.), Earth Rheology, Isostasy and Eustasy, Chichester: Wiley, 535-53.

Mörner, N.-A., 1984: 'Planetary, solar, atmospheric, hydrospheric and endogene processes as origin of climatic changes on the earth.' In N.-A. Mörner and W. Karlen (eds), Climatic Changes on a Yearly to Millennial Basis, Dordrecht: Reidel, 483-507.

Mörner, N.-A., 1987: 'Models of global sea-level changes.' In M. J. Tooley and I. Shennan (eds.). Sea-Level Changes, Oxford: Basil Blackwell Ltd., 332-55.

Newman, W.S. and Baeteman, C. 1987: 'Holocene excursions of the Northwest European Geoid.' Prog. Oceanog., 18, 287-322

Newman, W.S., Marcus, L.F., Pardi, R.R., Paccione, J.A. and Tomecek, S.M. 1980: 'Eustasy and deformation of the geoid: 1000-6000 radiocarbon years BP. In N.-A. Mörner (ed.) Earth Rheology, Isostasy and Eustasy.' Chichester: Wiley, 555-67.

Newman, W.S., Marcus, L.F. and Pardi, R.R. 1981: 'Palaeogeodesy, Late Quaternary geoidal configurations as determined by ancient sea levels.' In I. Allison (ed.) Sea Level, Ice and Climatic Change. Washington: IAHS Publ. No. 131, 263-75.

Peltier, W.R. 1982: 'Dynamics of Ice Age Earth.' Adv. Geophys., 24, 1-146.

Peltier, W.R. 1985: 'New constraints on transient lower mantle rheology and internal mantle bouyancy from glacial rebound data.' Nature, 318, 614-617.

Petersen, K.S. 1984: 'The Holocene marine history of the Limjford Area.' INQUA Subcommission on Shorelines of Northestern Europe Field Conference 1984, Septemnber 15-21. Excursion Guide, North Sea Coastal Zone between Jade Bay and Jammer Bight. Hannover: NLFB, 74-80.

Plassche, O. van de 1980: 'Holocene water-level changes in the Rhine-Meuse delta as a function of changes in relative sea level, local tidal range, and river gradient.' Geol. Mijnbouw, 59, 343-51.

Plassche, O. van de (ed). 1987: IGCP Project 200 Newsletter and Annual Report. Amsterdam: Free University.

Roeleveld, W. 1976: 'The Holocene evolution of the Groningen marine-clay district.' Berichten van de Rijksdienst voor het Oudheidkundig Bodemonderzock, vol. 24 Supplement.

Shennan, I. 1983: 'Flandrian and Late Devensian sea-level changes and crustal movements in England and Wales.' In D.E. Smith and A.G. Dawson (eds)., Shorelines and Isostasy, London: Academic Press, 255-83.

Shennan, I. 1986a: 'Flandrian sea-level changes in the Fenland, I: The geographical setting and evidence of relative sea-level changes.' J. Quaternary Science 1, 119-54.

Shennan, I. 1986b: 'Flandrian sea-level changes in the Fenland II: Tendencies of sea-level movement, altitudinal changes and local and regional factors.' J. Quaternary Science 1: 155-79.

Shennan, I. 1987a: 'Holocene sea-level changes in the North Sea Region.' In M.J. Tooley and I. Shennan (eds). Sea-Level Changes, Oxford: Basil Blackwell Ltd., 109-51.

Shennan, I. 1987b: 'Global analysis and correlation of sea-level data.' In R.J. Devoy (ed.), Sea-Surface Studies, Beckenham: Croom Helm, 198-230.

Shennan, I. 1987c: 'Impacts on the Wash of sea-level rise.' In P. Doody & B. Barnett (eds.), The Wash and its environment, Peterborough: Nature Conservancy Council, Research and survey in nature conservation series, 7, 77-90.

Shennan, I., Tooley, M.J., Davis, M.J. and Haggart, B.A. 1983: 'Analysis and interpretation of Holocene sea-level data.' Nature, 302, 404-6.

Sissons, J.B. 1967: The Evolution of Scotland's Scenery, Edinburgh: Oliver & Boyd.

Sissons, J.B. and Brooks, C.L. 1971: 'Dating of Early Postglacial Land and Sea Level Changes in the Western Forth Valley.' Nature, 234, 124-127.

Tooley, M.J. 1978: Sea-level Changes in North-west England during the Flandrian Stage, Oxford: Clarendon Press.

Tooley, M.J. 1982: 'Sea-level changes in Northern England.' Proc. Geol. Ass., 93, 43-51.

Tooley, M.J. 1987: 'Sea-level studies.' In M.J. Tooley and I. Shennan (eds.), Sea-Level Changes, Oxford: Basil Blackwell Ltd., 1-24.

ACKNOWLEDGEMENTS
Sincere thanks to Linda Scott for photographic reproductions, David Hume for cartographic work, and Louise Shennan for typing the manuscript.

SEA-LEVEL CHANGES IN THE IBERIAN PENINSULA DURING THE LAST 200,000 YEARS

C.Zazo, J.L.Goy
Dpto. Geodinámica
Fac. Geología
Universidad Complutense
28040 Madrid
Spain

ABSTRACT. A synthesis of the available data related to sea-level changes in the Iberian Peninsula (Spain) is done, from 200,000 yBP(first appearance of Strombus bubonius in the Mediterranean) up to now.

On the Atlantic coast, the best represented marine terrace is dated 100,000 yBP(Th/U), in the Cadiz littoral, and bears warm fauna such as Patella safiana, and it is affected by a strong deformation, the highest points being at the Gibraltar Strait. The Holocene deposits, dated 5,000 yBP, are widely distributed over the Cantabric coast and in Huelva. They often consist of systems of littoral spit bars with lagoons behind.

On the Mediterranean coast, four different marine terraces bearing Strombus bubonius have been found, and the ages, deduced from Th/U data, of the oldest three date from around 180 kyBP, 128 kyBP and 95 kyBP. All the terraces are tectonically deformed, presenting a general rising trend in the coast of Almería and sinking in the Mar Menor (Murcia) area; the area of Alicante again behaves as a rising zone, although to a lower degree than Almería.

INTRODUCTION

In this paper we will show the results of our own field research and of Th/U measurements, which have been determined in laboratories by research staff who have taken the samples together with us in the field.

The distribution of the recent Quaternary marine levels in Spain (Figure 1), takes place in a very unequal way all along the littoral, which comprises microtidal coasts, such as the Mediterranean, with a tidal range never higher than 0.5 m., and mesotidal coasts with a range of up to 4 m., such as in the Atlantic littorals. However, the fauna also changes from one basin to the other, Strombus bubonius

27

D. B. Scott et al. (eds.), Late Quaternary Sea-Level Correlation and Applications, 27–39.

28

being present in the Mediterranean and <u>Patella</u> <u>safiana</u> in the Atlantic.

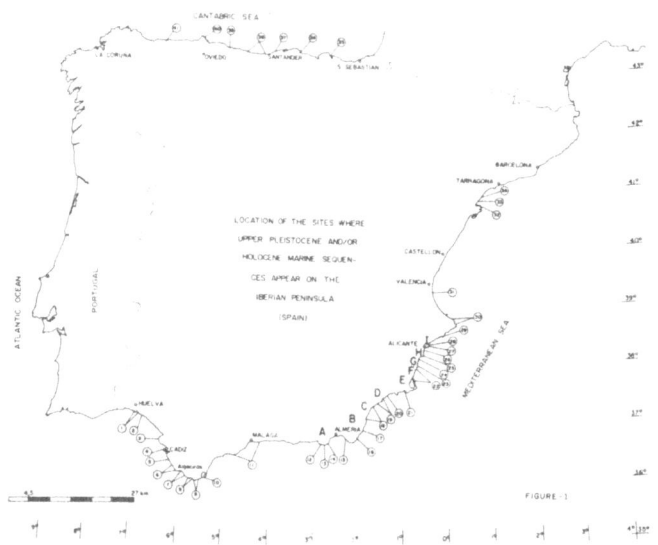

FIGURE 1

Much work has been carried out on the Spanish coasts related to the recent Quaternary strand lines, and, as it can be deduced from the synthesis done by the Spanish Working Group as a contribution to Project 200 of the I.G.C.P. (ZAZO, 1987), most data are concentrated around the Betic Mediterranean coasts. This area, and particularly the region of Almería, is where GOY and ZAZO (1982) defined, morphologically and stratigraphically, four marine episodes bearing <u>Strombus</u> <u>bubonius</u>, besides the Holocene level.

Several techniques have been used in order to date the marine terraces: Th/U, C^{14}, Pa/U, ESR and Amino Acids. The resulting dates do not always match the field data. This is due not only to the inherent problems of these techniques, and because fossil populations can be inherited from one terrace to the next, but also because samples are collected before careful and detailed geomorphological mapping is done and before sedimentologic criteris, such as plunge step sequence of the foreshore, location of the berms, etc., have been taken into account (BARDAJI et al., 1987).

I. ATLANTIC LITTORAL

I.1.Cantabric coast

The most outstanding morphological feature on this littoral

is a vast platform, partially of marine erosion, called "Rasa", where remains of deposits are sometimes preserved. It is dated upper Pliocene (HOYOS,1979) and it is 100-110 m. high at the most representative places.

The generally azoic marine levels from the upper Pleistocene appear encased in this surface, at a height of about 5-6 m. (the earliest one) and 2 m. (the latest one). These correspond to an estuarine level, widely represented in the Bañuges sector (Asturias)(HOYOS,1987), and dated by the presence of Acheulean industry.

The Holocene level consists of coastal barriers 1-1.5m. above mean high tide, and the present one appears attached to it. Some peat deposits, dated by C^{14} at 4,770 $\pm$ 110 yBP and 5,850 $\pm$ 130 yBP, have been found in some places on top of this Holocene deposit.

I.2.Huelva and Cadiz

From the upper Pliocene onwards, there is a generalized regression in this area which continues in the Quaternary, in such a way that the different oscillations of sea level in this period have left only small remains close to the present shoreline.

Likewise, during the Quaternary there is a change in the tectonic regimen, from an extensive phase that lasted from the upper Miocene, into a compressive one, which is noticeable by the change in the direction of ancient normal faults and by the setting of folds and faults. These events mainly affect the Quaternary marine levels, being responsible for their geomorphological disposition and spatial distribution (ZAZO,1980).

In the coast of Huelva, the Holocene deposits are extensively represented. During the maximum of the Flandrian transgression, the marshes were flooded by the sea; following a general SE longshore drift direction. The available C^{14} data give an age of 5,000 yBP for the beginning of the formation of these spit bars.

On the coast of Cadiz, the most continous marine terrace, which bears Thais haemastoma, Patella safiana and Cymatium doliarium, has been dated by Th/U and Pa/U at 90 $\pm$ $^{11}_{10}$ kyBP and 92 $\pm$ $^{8}_{4}$ kyBP (ZAZO,1980) and 105 kyBP (Th/U)(BRUCKNER and RADTKE,1986) Vast dune formations are always associated with these ancient beaches, and constitute the earliest Quaternary eolian complex preserved in the littoral of Cadiz. Following the most recent chronological scale established for the Moroccan coast /TEXIER et al.,1986)., this marine level corresponds to the Ouljian.

From a space/height diagram (Figure 2) of the transgressive maxima for this shoreline, a general rising trend can be deduced for this terrace in the area of the Gibraltar Strait, apart from the steep inflections produced by the passing of faults. The same phenomenon can be observed in the African coast (CADET et al., 1978), on the other side of the Strait.

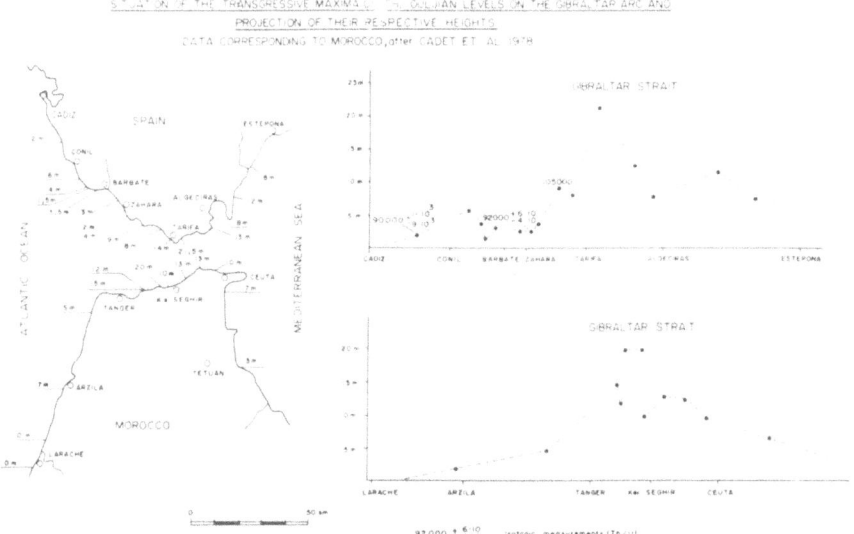

In the outskirts of Cádiz Bay, which behaves as a subsident area, the Quaternary marine levels are represented by lagoon-barrier island sequences, as at present.

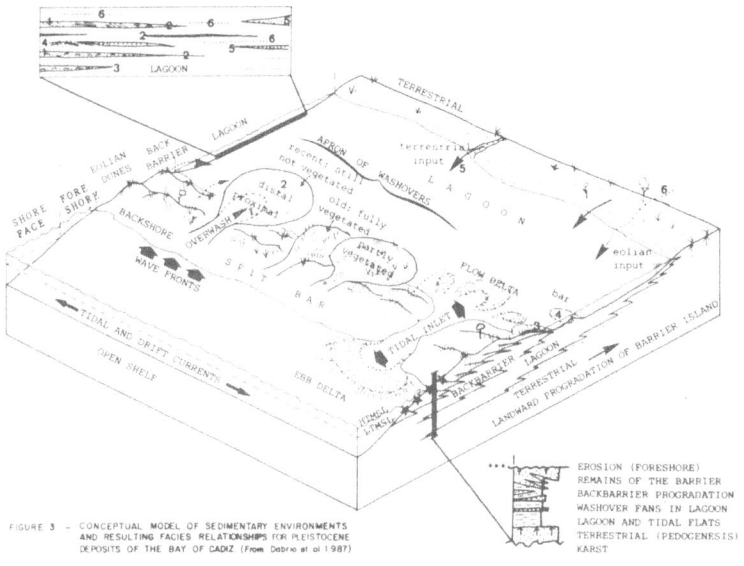

FIGURE 3 - CONCEPTUAL MODEL OF SEDIMENTARY ENVIRONMENTS AND RESULTING FACIES RELATIONSHIPS FOR PLEISTOCENE DEPOSITS OF THE BAY OF CADIZ (From Dabrio et al 1987)

The analysis of these sequences in the cliff of Puerto de Santa María, demonstrates the existence of at least three transgressive events, formed by lagoon-barrier island systems migrating landwards, and separated by erosional surfaces with karstic processes indicating falls of sea level.

The sedimentary model of the environment corresponding to these sequences is represented in <u>Figure 3</u>.

The chronology of the marine levels can not be precisely established, but the stratigraphy of the area suggests an Ouljian age for the latest unit (presence of <u>Thais haemastoma</u>, and dune cycle).

II. MEDITERRANEAN LITTORAL

This littoral constitutes the area where the highest number of Quaternary marine levels are represented, mainly in the Betic coast (Almería, Murcia and Alicante). The littoral of Valencia is poor in marine levels; besides the Flandrian, only one Quaternary paleobeach can be observed at +0.5 m. Its age could be middle Pleistocene or corresponding to the last Interglacial (GOY et al.,1987a), nevertheless we have never found <u>S.bubonius</u> in it.

More northwards, on the coast of Catalonia, near the Ebro delta, two overlapping marine levels bearing <u>S.bubonius</u> (PORTA and MARTINELL,1981) appear at +1 m. above sea level. For the latest one, BRUCKNER and RADTKE (1986) estimate 105 kyBP ± 20% and 137 kyBP ± 30% ESR ages. Due to the morphological and sedimentological characteristics of these two terraces, they have been correlated with the T.II and T.III units of Almería (ZAZO et al.,1981).

The main reason for the different distribution of the late Quaternary marine levels rests on the different tectonic behaviour of the different areas. While the Betic area has been subjected to a compressive regimen since the early Quaternary, the areas of Valencia and Catalonia have undergone an extensive phase since the Miocene.

The greatest number of observations and isotopic measurements have been carried out on the Betic littorals, not by chance, but because of the varied morphologic disposition and sedimentary environments of the Quaternary marine levels. More detailed description of these littorals is given below.

II.1Almería, Murcia and Alicante littorals.

The collision between the African and European plates is producing a compressive effect in these littorals, with a submeridian shortening axis (BOUSQUET, 1979) evidenced by the working of N120E and N140-160E dextral strike slip faults and N10E and N45E sinistral ones.

The Quaternary marine levels are affected by all these

32

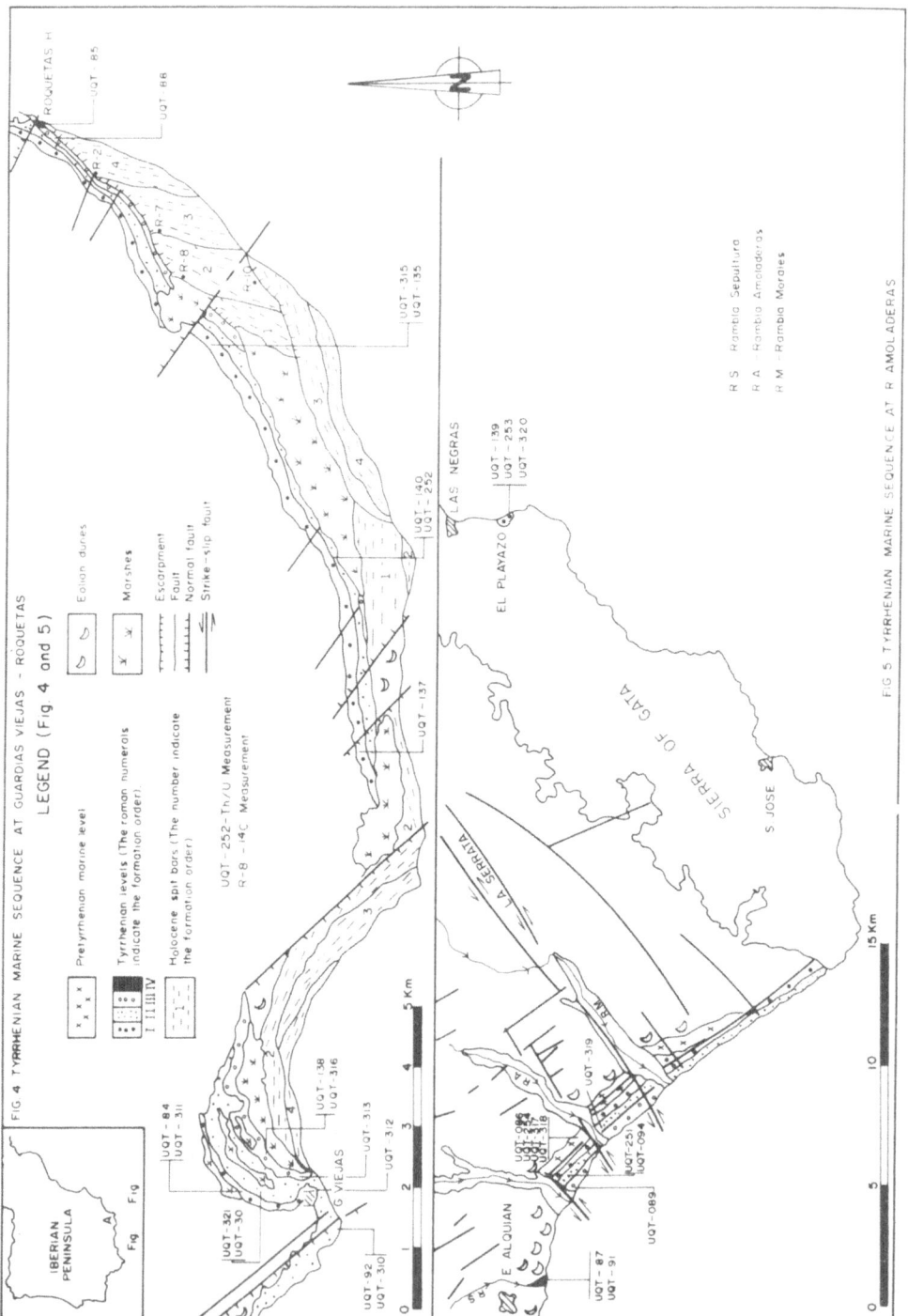

FIG 4 TYRRHENIAN MARINE SEQUENCE AT GUARDIAS VIEJAS - ROQUETAS

LEGEND (Fig 4 and 5)

Pretyrrhenian marine level

Tyrrhenian levels (The roman numerals indicate the formation order).

Holocene spit bars (The number indicate the formation order)

UQT-252 - Th/U Measurement
R-8 - 14C Measurement

Eolian dunes

Marshes

Escarpment
Fault
Normal fault
Strike-slip fault

FIG 5 TYRRHENIAN MARINE SEQUENCE AT R AMOLADERAS

R S   Rambla Sepultura
R A   Rambla Amoladeras
R M   Rambla Morales

IBERIAN PENINSULA

events (GOY et al.,1987b), which control the present height
of the transgressive maxima, the spatial distribution of the
levels, their geometric disposition with overlapping and/or
encasement, the morphology of the coastal outline, etc.

Four maine episodes, the deposits of which contain S.
bubonius (therefore Tyrrhenian in the sense of Issel), were
distinguished by GOY and ZAZO (1982) along the coast of Almería, and, later on, named T.I(the earliest), T.II, T.III
and T.IV(the latest)(ZAZO et al.,1984). Their distribution
has been successfully mapped in the areas of Guardias Viejas
(Figure 4) and Amoladeras (Figure 5). A detailed morphologic
and sedimentologic study indicates that these four units
each include minor oscillations of unknown significance (Figure 6).

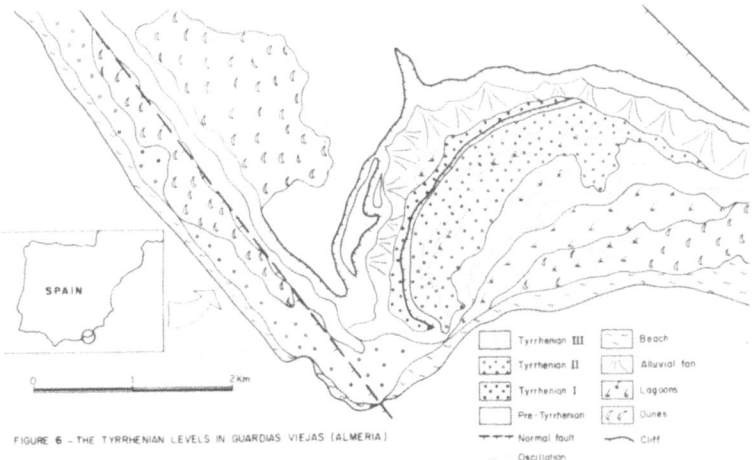

FIGURE 6 - THE TYRRHENIAN LEVELS IN GUARDIAS VIEJAS (ALMERIA)

From the geomorphologic point of view, the disposition
of the units is of encasement between the T.I and T.II and
T.III, which generally appear overlapping. The most recent
level, T.IV, has been observed in Rambla Sepultura (Almería)
where it seems to be overlapping the T.III, although this
can not be stated definitely due to the conditions of the
outcrop, and in Roquetas (Almería) where the T.IV appears
undoubtedly encased in the T.III (BAENA et al.,1982).

With respect to faunal distribution, the general rule
is that the T.I contains very few elements of S.bubonius,
while the T.II, T.III and T.IV become quite rich in this species. This trend is followed such that from Cabo de Gata (Almería) to Alicante we have never found S.bubonius in the morphostratigraphic unit T.I. although the accompanying warm
fauna is present.

33 Th/U ratios have been determined in molluscan fossils

TABLE I. Th/U MEASUREMENTS OF THE LITTORAL OF ALMERIA.

FIELD UNITS	SAMPLE NUMBER	AGE (years)
Pretyrrhenian	UQT-093..................>250.000	
	UQT-086..................>250.000	
	UQT-254..................>250.000	
	UQT-317..................>250.000	
	UQT-318..................>250.000	
T.I	UQT-84...................$159.000^{+16.800}_{-14.200}$	
	UQT-311.................$104.000^{+6.400}_{-6.000}$	
	UQT-137.................$175.500^{+20.000}_{-16.900}$	
	UQT-315.................$187.000^{+20.200}_{-17.000}$	
	UQT-135.................$188.400^{+22.400}_{-18.400}$	
	UQT-139.................$170.500^{+14.000}_{-12.400}$	
	UQT-253.................$188.600^{+28.000}_{-22.400}$	
	UQT-320.................$112.300^{+6.600}_{-6.200}$	
T.II	UQT-092.................$143.100^{+16.400}_{-14.000}$	
	UQT-310.................$132.800^{+12.000}_{-10.400}$	
	UQT-312.................$150.500^{+20.800}_{-17.200}$	
	UQT-140.................$130.600^{+12.800}_{-11.400}$	
	UQT-252.................$131.100^{+14.000}_{-12.200}$	
	UQT-251.................$201.300^{+30.600}_{-24.900}$	
	UQT-094.................$131.200^{+10.600}_{-9.600}$	
	UQT-319.................$132.900^{+12.200}_{-10.800}$	
	UQT-090.................$150.300^{+12.600}_{-12.000}$	
	UQT-321................ >250.000	
T.III	UQT-313.................$195.000^{+28.000}_{-22.000}$	
	UQT-138.................$190.000^{+24.200}_{-20.400}$	
	UQT-316.................>250.000	
	UQT-085.................$100.600^{+9.000}_{-8.400}$	
	UQT-091.................$92.600^{+6.400}_{-5.400}$	
T.IV	UQT-88..................>250.000	
	UQT-87..................$32.200^{+3.000}_{-3.000}$	
	UQT-91..................$55.600^{+3.000}_{-3.000}$	

from the various units in Almería (HILLAIRE-MARCEL et al.,
1986). A histogram of Th/U values from the three older units
"clustered reasonably around values of about 0.85(~180kyBP),
0.71(~128kyBP) and 0.59(~95kyBP)"(ibid, p.61 and Fig.2).

The location of the samples is shown in Figures 4 and 5.
In Table I the field units are contrasted with the results
of the isotopic measurements made in them. The most anomalous results occur in the episode T.IV. The samples taken
from this unit are suspected of being a geochemically open
system (the ages range from 32 to 250 kyBP). Further measure-
ments(ESR-spectra), carried out in the mouth of Rambla Que-
brada (BRUCKNER,1986) beside Rambla Sepultura, have given a
Holocene age for this deposit. For us, this unit could never
be Holocene because of field evidence; it seems to be the
same as the one at Roquetas, where the exposed section shows
clearly the direct relation between episode IV and the other
Tyrrhenian terraces, and where that deposit is strongly eroded by the Holocene spit-bar system (maximum C^{14} age 6,500yBP),
Table II (GOY et al., 1986).

TABLE II. C^{14} MEASUREMENTS OF THE LITTORAL OF ALME-
RIA (spit-bars of Roquetas),(GOY et al.,1986)

Sample	Age (y.)-(Libby)	Spit-bar systems
R-8	6,450 ± 100	2
R-7	3,600 ± 100	2
R-10	2,150 ± 400	3
R-2	1,870 ± 35	4

Likewise, the fauna found in these spit-bars is exactly
the same as the present Mediterranean one, while the deposit
considered as T.IV bears not only S.bubonius (which could be
inherited) but also some corals and Lithotamnion algae.

Much work on geomorphological mapping and detailed sedimentologic and neotectonic studies has been carried out in
the littorals of Murcia (BARDAJI et al.,1986) and Alicante
(GOY et al.,1987c; GOY and ZAZO,1987), where sampling for
Th/U dating has been done.(The results are not yet published).
From this, it can be deduced that the field units T.I,T.II
and T.III, defined in the area of Almería (Guardias Viejas-
Roquetas,Amoladeras,Mojácar,Pozo del Esparto, Figure 1:A,B,C
and D respectively) are localized in those areas with a rising trend on the littoral of Alicante (La Marina and Santa Po-
la, Figure 1:G and H). While in the zones of Murcia (Mar Menor,Figure 1:E) and Alicante (Torrevieja-La Mata lagoons, Figure 1:F), the sinking trend during the Quaternary makes the
Tyrrhenian sequence appear discontinous and the morphological

36

disposition of the levels is overlapping or superposition
instead of encasement.

The episode T.IV, defined in Almería, seems to be represented again in Cabo de las Huertas (Alicante)(Figure 1:
I). On the outermost tip of the cape, a beach bearing S.bu-
bonius appears at +2 m., which has been dated 91 ± 6 kyBP
(DUMAS,1977) and 85 ± 5 kyBP (STEARNS and THURBER,1965)
placing it in the T.III unit. Another marine level appears
2 km. westwards at about +3 m., with a vast number of S.bu-
bonius. The sedimentologic characteristics (cementation degree) suggest a younger age than the former one. However,
the isotopic measurements (Th/U) carried out on these mollusca, display contradictory results: 32 ± 3 kyBP (STEARNS and
THURBER, 1965); 38 ± 2 kyBP (DUMAS,1977) and 191 ± 5 kyBP
(HILLAIRE-MARCEL et al., 1986,sample UQT-083).

Figure 7 shows schematically the tectonic trend produced by N-S Quaternary compression on the coasts of Almería,
Murcia and Alicante. The field data demonstrate a southward
shifting of the sinking area from Tyrrhenian I to Tyrrhenian
III (SOMOZA et al.,1987).

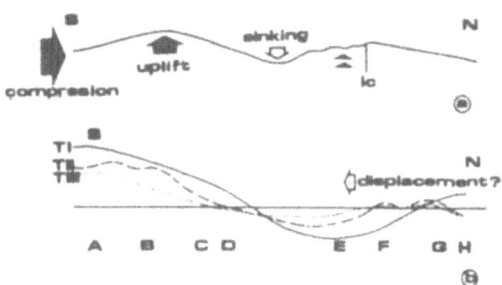

FIGURE 7 - Regional Tectonic trend
        a) The Quaternary compression S-N provokes a high
           range deformation. Uplift in the south area
           (Almería), sinking in the central zone (Mar Me
           nor) and light uplifting near the Crevillente
           limit (lc)
        b) General tectonic trend for each episode (T.I,
           T.II,T.III) showing the southward displacement
           of the negative trend zone. A-H indicate the
           points located in figure 1.
           (After SOMOZA et al.,1987)

CONCLUSIONS

The analysis of the sea level changes for the last 200,000
years along the coasts of the Iberian Peninsula (Spain), reveals an unequal distribution along the littoral, mainly due
to differences in the geodynamic behaviour of the coast.

Hence, the areas with the largest number of remains of ancient sea levels are the areas close to the boundary of plates (Mediterranean coast, and littorals of Almería, Murcia and Alicante).

On the Atlantic coast, the best developed level is the one of about 100 kyBP, on the littoral of Cadiz, where tectonics have affected it in such a way that it appears rising in the Gibraltar Strait area, as in the African coast, and sinking toward the Atlantic and the Mediterranean.

The remains of Holocene beaches (coast of Asturias and Huelva) are developed at a slightly greater height than present sea level, and are dated around 5,000 yBP.

On the Mediterranean coast, four marine terraces bearing Strombus bubonius (i.e. Tyrrhenian), can be morphologically and sedimentologically distinguished: T.I (the earliest), T.II, T.III and T.IV (the latest). Isotopic measurements (Th/U) have been carried out on these terraces in the littoral of Almería, with the resulting ages of 180 kyBP(T.I) 128 kyBP(T.II) and 95 kyBP(T.III). For the most recent one (T.IV) the results are contradictory.

The morphological disposition of these units is encasement in the areas with rising trends, and overlapping or superposition in zones of sinking trends.

38

REFERENCES

BAENA,J.,GOY,J.L.,ZAZO,C.(1981).'Litoral de Almería'. En:
Excursión-mesa redonda sobre el Tirreniense del litoral me-
diterráneo español.Libro Guía. Madrid-Lyon,25-44.
BARDAJI,T.,CIVIS,J.,DABRIO,C.J.,GOY,J.L.,SOMOZA,L.,ZAZO,C.
(1986).'Geomorfología y estratigrafía de las secuencias marinas y continentales cuaternarias de la cuenca de Cope(Mur
cia,España)'.En: Estudios sobre Geomorfología del Sur de Es
paña. Ed.   F.LOPEZ BERMUDEZ;J.B.THORNES.Dep.Geogr.Fisica,
Univ. de Murcia,11-16.
BARDAJI,T.,DABRIO,C.J.,GOY,J.L.,SOMOZA,L.,ZAZO,C.(1987).'Sedimentologic features related to Pleistocene sea level changes in   SE Spain'. In: Late Quaternary Sea Level Changes
in Spain, Ed   C.ZAZO. C.S.I.C.Madrid, 79-93.
BOUSQUET,J.C. (1979). 'Quaternary strike-slip faults in
southeastern Spain'. Tectonophysics,53,274-286.
BRUCKNER,H.(1986).'Stratigraphy, evolution and age of Quaternary marine terraces in Morocco and Spain'.Z.Geomorph.N.
F.,62,83-101.
BRUCKNER,H.,RADTKE,U.(1986).'Paleoclimate implications derived from profiles along the Spanish Mediterranean coasts'.
In: Quaternary Climate in Western Mediterranean, Ed. F.LO-
PEZ VERA. Univ. Autónoma de Madrid,468-486.
CADET,J.P.,FOURNIGUET,J.,GIGOUT,M.,GUILLEMIN,M.,PIERRE,G.
(1978).'La néotectonique des littoraux de l'arc de Gibral-
tar et des partout de la mer d'Alboran', Quaternaria,XX,
185-202.
DUMAS,B.(1977). 'Le Levant espagnol. La genése du relief'.
Thése d'Etat, Univ. Paris XII,520p.
GOY,J.L.,ZAZO,C.(1982).'Niveles marinos cúaternarios y su
relación con la tectónica en el litoral de Almería'.Bol.R.
Soc.Esp.Hist.Nat.(Geol.),80,171-184.
GOY,J.L.,ZAZO,C.,DABRIO,C.J.,HILLAIRE-MARCEL,C. (1986).'Evo
lution des systemes de lagoons-isles barriéres du Tyrrheni-
en á l'actualité á Campo de Dalias(Almería,Espagne). In:
Changements Globaux en Afrique durant le Quaternaire.Passé-
Présent-Future. Ed.   Orstom.Paris,197,169-172.
GOY,J.L.,ZAZO,C.(1987).'Quaternary shorelines and their dis
position related to the continental deposits and neotectonics in the Elche depression(Alicante,Spain)'. In: Programme with abstracts, INQUA XII International Congress,Canada,
176.
GOY,J.L.,REY,J.,DIAZ DEL RIO,V.,ZAZO,C.(1987a).'Relación en
tre las unidades geomorfológicas cuaternarias del litoral y
de la plataforma interna-media de Valencia,(España).Implica
ciones paleogeográficas.' Com.III Reun.Nac.G.A.O.T., Univ.
de Valencia,II,1369-1381.
GOY,J.L.,ZAZO,C.,BARDAJI,T.,SOMOZA,L.(1987b)'Las terrazas
marinas del Cuaternario reciente en los litorales de Murcia
y Almería(España): El control de la neotectónica en la dis-

posición y número de las mismas'.Est.Geológicos,42,439-443.
GOY,J.L.,SOMOZA,L.,BARDAJI,T.,ZAZO,C.(1987c).'Shoreline map
ping models in areas with different morphosedimentary beha-
viour(Almería,Murcia, Alicante,Spain) In: Late Quaternary
Sea-Level Changes in Spain, Ed. C.ZAZO.C.S.I.C.Madrid,35-
47.
HILLAIRE-MARCEL,C. ,CARRO,O.CAUSSE,C. ,GOY,J.L.,ZAZO,C.
(1986). 'Th/U dating on Strombus bubonius bearing marine terraces in southern Spain'. Geology,14,613-616.
HOYOS,M.(1979).'El Karst en Asturias durante el Pleistoceno
superior y Holoceno'.Ph.Thesis.Univ.Complutense de Madrid,
385 p.
HOYOS,M.(1987).'Upper Pleistocene and Holocene marine level
on the Cornisa Cantábrica (Asturias,Cantabria and Basque
Country)Spain'. In: Late Quaternary Sea-Level Changes in
Spain, Ed. C.Zazo, C.S.I.C.,Madrid,251-258.
PORTA,J.,MARTINELL,J.(1981).'El Tirreniense catalán, sínte
sis y nuevas aportaciones'. Publ.Dep.de Paleontología,Uni-
versitat de Barcelona,3-27.
SOMOZA,L.,ZAZO,C.,BARDAJI,T.,GOY,J.L.,DABRIO,C.J.,(1987).
'Recent Quaternary sea-level changes and tectonic movement
in SW Spanish coast'. In: Late Quaternary Sea Level Changes
in Spain, Ed. C.ZAZO,C.S.I.C. Madrid,49-77
STEARNS C.E.,THURBER,D.L.(1965).'Th230-U234 dates of late
Pleistocene marine fossils from the Mediterranean and Moroc-
can littorals'. Quaternaria,VII,29-42.
TEXIER,J.P.,RAYNAL,J.P.,LEFEVRE,D.C.(1986).'Thoughts on the
Quaternary chronology of Morocco'. In: Quaternary Climate
in Western Mediterranean, Ed. F.LOPEZ VERA,Univ.Autónoma
de Madrid,487-502.
ZAZO,C.(1980).'El Cuaternario marino-continental y el lími
te Plio-Pleistoceno en el litoral de Cádiz. Ph.Thesis, Univ
Complutense de Madrid, 2 Vol.
ZAZO,C.,GOY,J.L.,HOYOS,M.,DUMAS,B.,PORTA,J.,MARTINELL,J.,
BAENA,J.,AGUIRRE,E.(1981).'Ensayo de síntesis sobre el Ti-
rreniense peninsular español'.Est.Geológicos,37,257-262.
ZAZO,C.,GOY,J.L.,AGUIRRE,E.(1984).'Did Strombus survive the
last Interglacial in the Western Mediterranean?',Mediterra-
nea,3,131-137,
ZAZO,C.(Ed.), (1987).Late Quaternary Sea Level Changes in
Spain. Trab.Neogeno-Cuaternario,10,303 p.

This work has been supported by the Spanish CAICYT,Project
2460/83.

SEA-LEVEL CHANGES IN THE NETHERLANDS DURING THE LAST 6500 YEARS: BASAL PEAT VS. COASTAL BARRIER DATA

Orson van de Plassche & Thomas B. Roep
Institute for Earth Sciences
Free University
P.O. Box 7161
1007 MC Amsterdam
The Netherlands

ABSTRACT. This paper compares a recently published Mean Sea Level (MSL) curve, derived from paleo-tide evidence recorded in coastal barrier deposits in the western Netherlands, with the latest of three MSL curves published since 1961, all based on indicators that originated at various distances from the contemporaneous open sea or coastline, predominantly basal organic deposits. The two MSL curves compare very well. To establish whether this good agreement is real or apparent, the accuracy of the two curves is discussed. An error band for the upper limit of MSL is derived from available basal peat data, while a lower limit is obtained from various other kinds of back barrier sea-level evidence. Although no error envelope is established for the coastal barrier MSL curve, it can be concluded that the agreement between both MSL curves is, for the most part, likely to be real.

1. INTRODUCTION

Recently, Roep and Beets (1988) summarized results of studies on paleo-tide levels begun in 1971 and derived from sedimentary characteristics of coastal barrier deposits in the western Netherlands. The MSL, or rather the Half Tide Level (HTL) curve, they inferred from their data (hereafter referred to as the cb-MSL curve) compares very well with an earlier MSL curve (called bp-MSL), which is primarily based on dated basal peat samples (Fig. 1) collected mainly by Jelgersma (1961, 1977, 1979) and Van de Plassche (1982). The remarkable similarity between the two MSL curves (tide markers from the open coast vs. groundwater indicators that originated many tens of kilometers from the contemporaneous littoral environment) raises the question of the accuracy of both curves.

D. B. Scott et al. (eds.), Late Quaternary Sea-Level Correlation and Applications, 41–56.
© *1989 by Kluwer Academic Publishers.*

Both the bp- and cb-MSL curve are considered accurate to within
+/- a few decimeters. In neither case, however, is this semi-
quantitative value based on a systematic evaluation of the vertical
errors involved. Following a concise review of sea-level studies in
the Netherlands since 1953, the question of the error envelopes for
both MSL curves is discussed, though in this paper one is established
for the bp-MSL curve only. A brief discussion of ongoing and planned
sea-level research activities in the Netherlands concludes the paper.

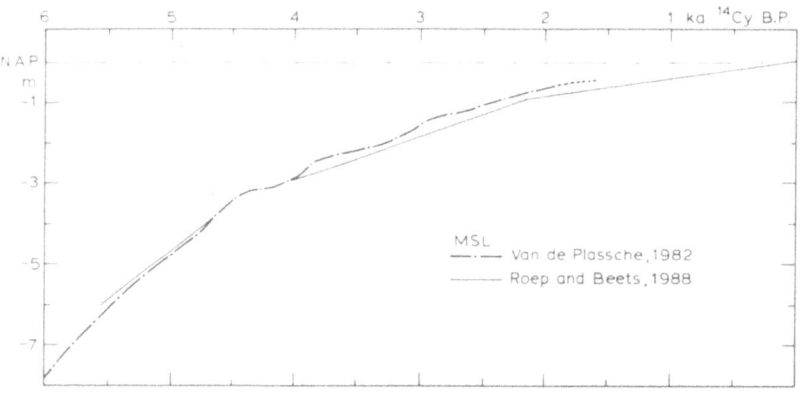

Figure 1.   Comparison of MSL curves based on basal peat data (Van de
Plassche, 1982) and on coastal barrier data (Roep and Beets, 1988).

2.   SEA-LEVEL STUDIES BETWEEN 1953 AND 1983

Between 1953 and 1983, several modified MHW and MSL curves, each based
on different or strongly enlarged sets of data, were published for
parts of the Netherlands (Van Straaten, 1954; Bennema, 1954;
Jelgersma, 1961, 1966, 1977, 1979, 1980; Louwe Kooijmans, 1974, 1976;
Roeleveld, 1974; Van de Plassche, 1982).

The first paper on sea-level changes in the Netherlands, after
the radiocarbon method was introduced, deals with sedimentary
structures in coastal deposits as indicators of former sea-level
positions (Van Straaten, 1954). He analysed tidal deposits in a deep
tunnel pit at Velzen (Fig. 2). He realized that the presence of peat
and peaty clay layers in the sequence rendered the site unsuitable for
accurate sea-level study because of sediment compaction. However, his
impressive attempt to do so is understandable because of his wide
knowledge of the tidal flat environments and because, for the first
time in the Netherlands, a series of radiocarbon dates on shells from
marine influenced deposits at one locality became available.

In the method employed by Van Straaten, the indicative value of
the sedimentary structures was clear and relatively accurate, but

compaction resulted in large error margins. In the approach taken by others thereafter, especially Jelgersma (1961), the compaction problem was eliminated or small, but the relationship to sea level of groundwater indicators used, primarily dated basal peat samples, remained uncertain for a long time.

Jelgersma (1961) collected basal peat samples (1) in the Province of Zeeland, (2) from the slopes of two almost completely submerged, late glacial river dunes in the Rhine/Meuse delta, and (3) at various sites in the Provinces of South and North Holland, Friesland and Groningen. Each of these three data sets gave rise to a separate curve (Fig. 3), although the "river dune" data contributed to the construction of what became known as the Jelgersma curve (curve 1 in Fig. 3). The higher position of the Zeeland curve was tentatively explained in terms of a larger tidal range and/or less tectonic subsidence due to proximity to the area of the Brabant Massif.

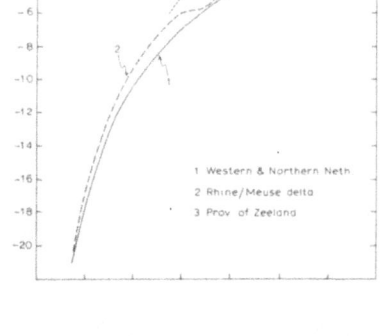

Fig. 2. Location map of the Netherlands with topographic names mentioned in the text. Abbreviations: A - Alkmaar; F - Friesland; G - Groningen;

Fig. 3. Time-depth graphs, based on basal peat data, by Jelgersma (1961).

H - Haarlem; Me - Meuse; N-H - North-Holland; R - Rotterdam; Rh- Rhine; ORh - Older Rhine; S-H - South-Holland; TH - The Hague; V- Velzen; Z - Zeeland (after H. Harlan Molenaarsgraaf).

Jelgersma's curve for the western and northern Netherlands has been modified several times (Fig. 4). The first change concerned a correction for the Suess effect (Vogel and Waterbolk, 1963). The corrected curve was first published in Pons et al. (1963), but is more often referred to in Jelgersma (1966). The second modification concerned a change in meaning of the curve, which was believed to indicate the rise of MHW. In 1975, the first results were announced of a study on paleo-tide levels derived from shell-dated sedimentary structures in the coastal barrier deposits of the western Netherlands (Roep et al., 1975; Jelgersma et al., 1975). The time-depth position of the paleo-MHW levels obtained appeared to be about 0.8 to 1 m above Jelgersma's curve, a value equal to or slightly larger than the present-day tidal amplitude along the western coastline. In light of this important new evidence, Jelgersma's curve, or at least the part younger than 5000 yBP, was reinterpreted to indicate the rise of MSL rather than of MHW. Finally, Jelgersma (1977, 1979) steepened the older part of her curve on the basis of new data from the North Sea and a new basal peat date from central North Holland.

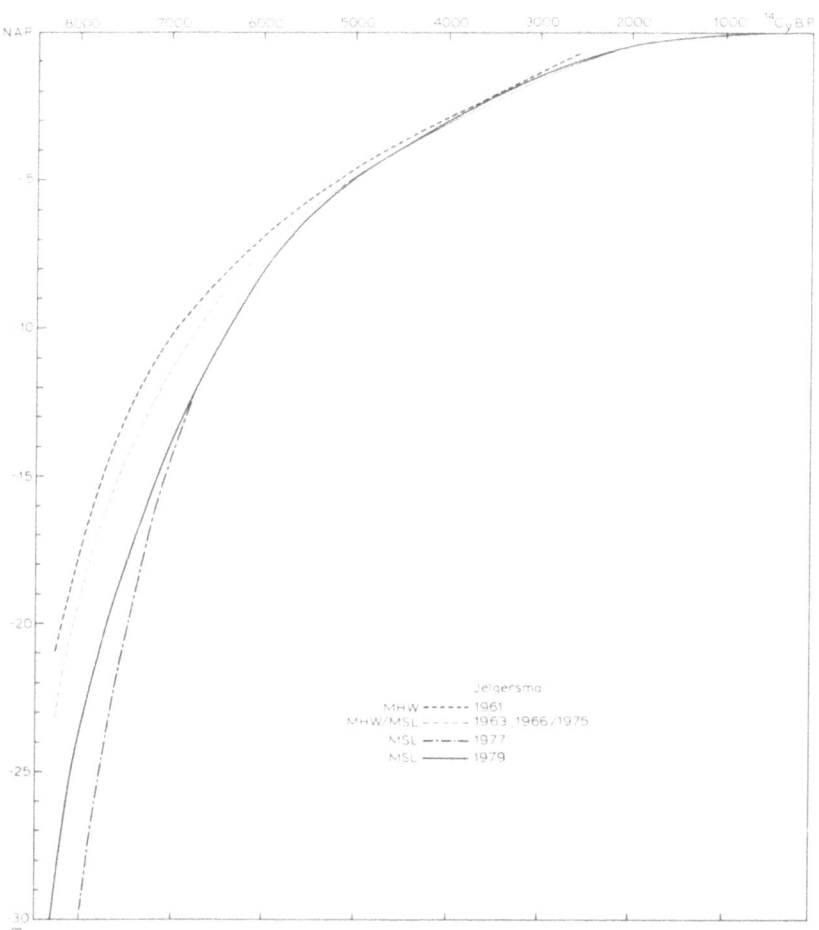

Figure 4.   Modified versions, by Jelgersma, of her MHW/MSL curves (from Van de Plassche, 1982).

An element of competition entered the scene of sea-level studies in the Netherlands when Louwe Kooijmans (1974) published a fluctuating (stepped) MHW curve for the western part of the country. This curve is based almost entirely on estimates of former local MHW or groundwater levels at archaeological sites (c.f. Bennema 1954). And when, in the same year and according to a completely different method, Roeleveld (1974) produced a smooth MHW curve for the northern coastal district that strongly resembled the trend of Louwe Kooijmans' (1974) curve (Fig. 5), it seemed that Jelgersma's graph would have to be put to discussion. However, although one year later, Jelgersma et al. (1975) would announce that her curve represented the rise of MSL rather than of MHW, it was Louwe Kooijmans' idealized, interpretative, stepped curve that evoked more discussion. He reacted to this with a methodologically important paper (Louwe Kooijmans, 1976) in which he gave a new, smooth curve for the rise of MSL complete with error margins, based on a selection of his most suitable data (Fig. 6). The fact that the part of Louwe Kooijmans' MSL curve older than 3000 yBP occurs above that of Jelgersma (1979) can be attributed to an underestimation of the river-gradient effect for the older time-depth data from the Hazendonk river dune near Molenaarsgraaf in the central part of the Rhine/Meuse delta, and the effect of a groundwater gradient on the position of his oldest data point (Van de Plassche 1982).

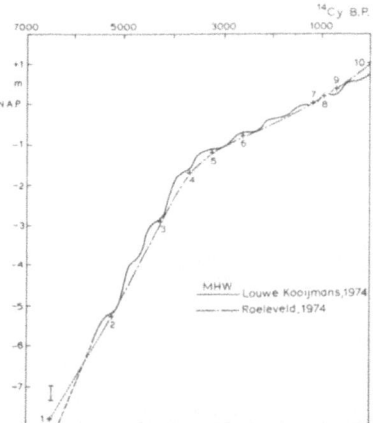

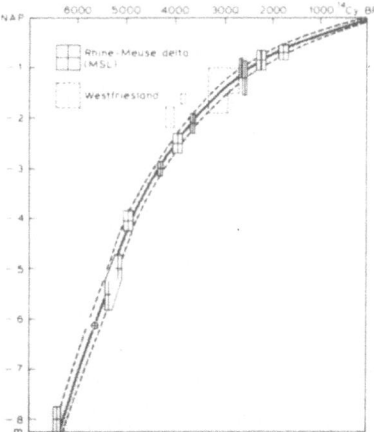

Fig. 5. Comparison of the MHW curves by Louwe Kooijmans (1974; western Netherlands) and Roeleveld (1974, northern coastal district; not corrected for compaction) (modified after Roeleveld, 1974).

Fig. 6. MSL curve for the western Netherlands by Louwe Kooijmans (1976) (slightly modified after Van de Plassche, 1982).

The MHW curve of Roeleveld (1974)(Fig. 5) is a by-product of his stratigraphic-analytical approach to the coastal sedimentary sequences encountered in the Province of Groningen. Because the compaction factor could not be eliminated (or assumed negligible), the true MHW curve for the Groningen coastal area should lie slightly above Roeleveld's curve, except for the oldest part of his curve which is too high due to the fact that the indirectly derived age of his oldest/deepest time-depth point (ca. 6500 yBP) is about 800 to 900 radiocarbon years too old (Van de Plassche, 1982 and unpublished radiocarbon data).

Furthermore, analysis of the time-depth data from the northern Netherlands also shows that Roeleveld's conclusion that the northern coastal district may be subject to less crustal subsidence than the western coastal area cannot be maintained. Recently, Shennan (1987) concluded that the Groningen coastal area may have experienced slightly greater subsidence than the western Netherlands. However, the fact that two basal peat dates from NE Friesland (ca. 5800 yBP and 4700 yBP; Griede 1978) coincide with Jelgersma's (1979) basal peat curve indicates that for the last 6000 radiocarbon years differential tectonic and/or isostatic movements between the northern and the western Netherlands have been too small to detect (Van de Plassche, 1982, Fig. 28).

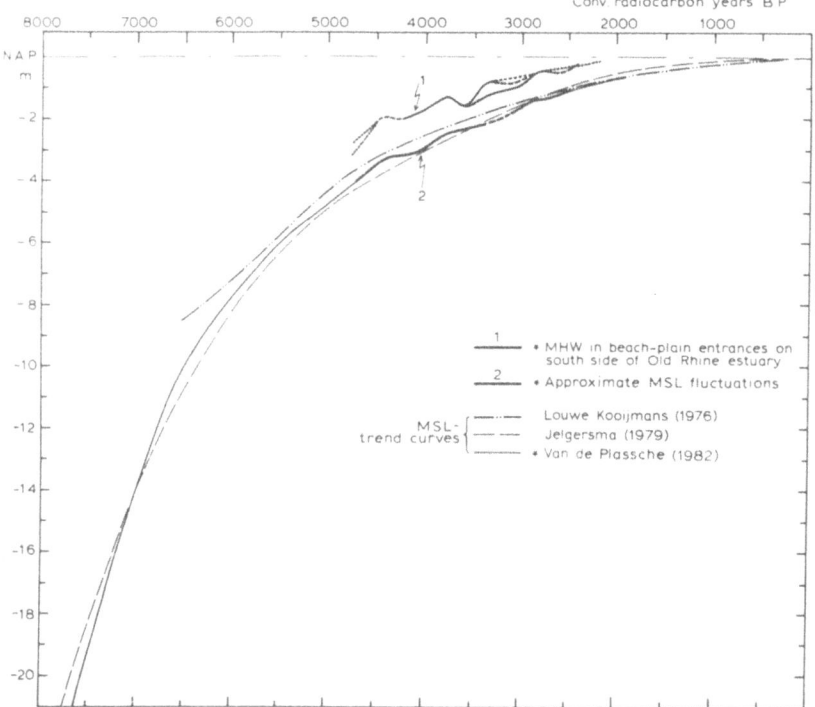

Figure 7. Sea-level curves (MSL, local MHW) by Van de Plassche (1982) compared to MSL curves by Louwe Kooijmans (1976) and Jelgersma (1979) (from Van de Plassche, 1982).

The data and methods used by Jelgersma, Louwe Kooijmans and Roeleveld, and some early and recent data by Van Straaten (1954), Bennema (1954) and Griede (1978) were all discussed and evaluated in detail by Van de Plassche (1979, 1980, 1982). In addition, building further on the approach and results of Jelgersma (1961), Van de Plassche (1982) collected ca. 60 new basal peat samples from two river dunes near Rotterdam and from the beach-plain entrances bordering the former Older Rhine estuary on the south. The curves based on these samples and a selection of earlier data, are shown in Figure 7. The fluctuations in the MSL curve between 4600 and 2500 yBP (curve 2, the bp-MSL curve) are interpreted. They are based primarily on the observation that fluctuations in the local water-level rise, as suggested by the basal peat time-depth data from the river dunes near Rotterdam (including the Barendrecht river dune sampled by Jelgersma, 1961) are closely paralleled by fluctuations in the rise of the local MHW level that can be shown or inferred to have occurred in the beach-plain entrances more than 30 km to the NNW.

3.   NEW SEA-LEVEL DATA SINCE 1982

Recently, a completely new and independently obtained set of smooth sea-level curves were published by Roep and Beets (1988). In Figure 8, these curves have been very slightly modified to correct for some small technical drawing errors. The curves approximate the rise of MHW, MLW and inferred MSL (=HTL) <u>along the open coast</u> of the western Netherlands, because they are entirely based on indicators of paleo-tide levels recorded in coastal barrier and other littoral deposits between Alkmaar and The Hague (Roep et al., 1975; Roep et al., 1979; Beets and Roep, 1981; Roep et al., 1983; Van der Valk et al., 1985).
The MHW curve is derived from the deepest occurrence of dry eolian scour and the highest levels of marine burrows and small-scale cross-lamination, while the MLW curve is based on the level of thickest shell beds, the vertical range of structureless sand, bubble sand and/or low-angle bars and the occurrence of cm-thick clay intercalations. The cb-MSL (HTL) curve is drawn midway between the MHW and MLW curves. For the oldest part, however, the position of the curve was determined by considering the 3 radiocarbon dates and related sedimentary characteristics as 1 unit, and this explains why the oldest part of the bc-MSL curve in Fig. 8 is slightly higher than the same part of the bc-MSL curve in Figs. 1 and 9 (radiocarbon time scale), where it is derived strictly as a HTL curve from the given MHW and MLW curves. The youngest part is controlled by modern tide-gauge data. This data set is too small to detect small-scale changes in rate of sea-level rise, but does seem to indicate that tidal range during the last 2000 years was somewhat smaller than between 4500 and 3000 cal BC.

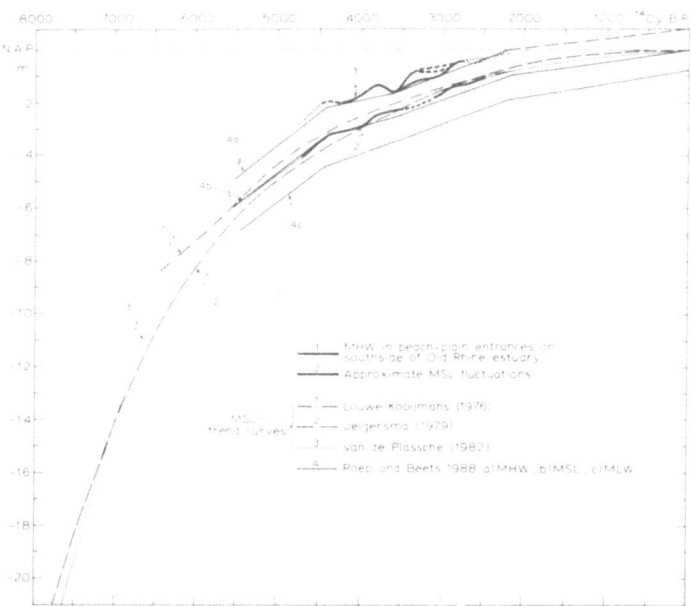

Figure 8. Comparison of selected MHW and MSL curves published for the western (and northern) Netherlands based on (base of) basal peat data (curves 2 and 3), other back barrier data (curve 1), beach-plain peat data (curve 1 fat) and coastal **barrier** data (curves 4a, b and c).

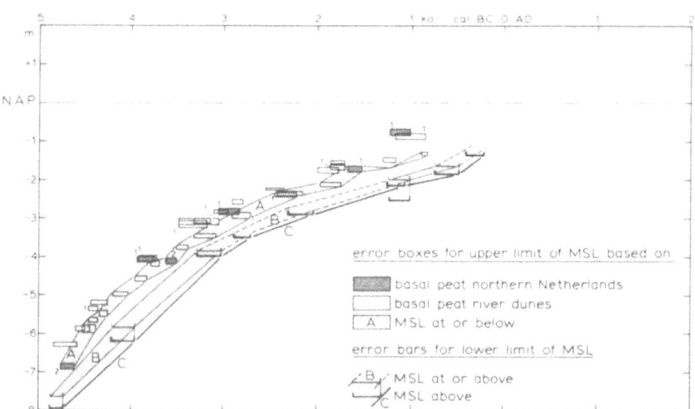

Figure 9. Time-depth diagram showing error bands for the upper and lower limit of the MSL rise between 5000 and 500 cal BC. Error band A is based on all basal peat data younger than 6000 BP used in the construction of the bp-MSL curve in **Fig.** 7. 1) Jelgersma, 1961; 2) Griede, 1978; 3) Ente, 1976; and unnumbered boxes: Van de Plassche, 1982.

4.   BASAL PEAT VS. COASTAL BARRIER MSL CURVES

As shown in Fig. 8, the cb-MSL curve by Roep and Beets (1988) compares
well with all of the earlier MSL curves, in particular with the latest
of these (curve 3:   the bp-MSL curve), which is built on basal peat
data obtained by Jelgersma (1961, 1977), Ente (1976), Griede (1978)
and Van de Plassche (1982).   Given the completely different nature of
the data, this is an important result, since if it can be shown that
both the bp- and cb-MSL curves are reasonably accurate, it lends
support to the assumptions, interpretations and estimates made.

4.1   Accuracy of the bp-MSL Curve

4.1.1 Qualitative Assessment.   Before discussing the error margins of
the bp-MSL curve (curve 3a in Fig. 8) quantitatively, it should be
noted that there is reason to expect that the curve probably is, apart
from the small, interpreted fluctuations, a rather good approximation
of the MSL rise in the western and northern Netherlands.   This may be
concluded from Fig. 65 in Van de Plassche (1982), in which the
difference and variation in time-depth position of 3 independently
established,   smooth   curves   can   be   given   a   very   satisfactory
interpretation.   The   first   curve,   which   is   based   on   a   critical
evaluation   of   sea-level   data   published   between   1953   and   1980,
approximates the rise of MSL.   The second curve, built on almost 40
basal peat samples from river dunes in the Rhine/Meuse delta, lies
above the MSL curve (ca. 2 m around 7000 yBP), but gradually converges
on, and as of ca. 5300 yBP, coincides with it.   A third curve, derived
from beach-plain peat data and covering the period 4500-2300 yBP, more
or less parallels (slightly converges on) the corresponding parts of
the two coincident curves at a level between 1 and 0.8 m higher, which
is very close to modern values for the tidal amplitude along the
western Netherlands coastline (0.8 ± 0.15 m).
       Convergence of the river-dune basal peat curve with the MSL curve
can be fully explained in terms of a decreasing river-gradient effect
in the delta, and an increasing, and ultimately complete, damping of
the tidal range in the vicinity of the sampled river dunes (near
Rotterdam,   many   tens   of   kilometers   away   from   the   contemporaneous
coastline).   The   fact that the "river dune" curve nowhere dives below
the MSL curve and that the "beach plain" curve, which can be expected
to approximate the rise of a water level between coastal MHW and MHWS
(Spring) level, occurs at a very reasonable height above it, may be
considered   as   consistent   indications   that   the   bp-MSL   curve   is   a
reliable approximation of the true MSL rise.

4.1.2 Quantitative Assessment.   Figure 9 is a time-depth diagram
showing error bands for the upper and lower limit of the MSL rise
between 5000 and 500 cal BC.   Error band A is based on all basal peat
data younger than 6000 yBP used in the construction of the bp-MSL
curve in Fig. 7 (curve 3 in Fig. 8), while band B is derived from a
variety of other kinds of sea-level indicators.   Radiocarbon ages have

been calibrated according to Pearson et al. (1986).

4.2   Upper Limit for MSL

Band A (Fig. 9) was constructed by 1) taking the measured depth and
sample thickness of all basal peat data younger than 6000 yBP (ca.
5000 cal BC), both from the river dunes in the Rhine/Meuse delta and
from the Pleistocene subsurface in the northern Netherlands, 2)
multiplying the sample thickness by a factor of 2 (to account for
compaction), 3) adding error margins for measuring the depth of a
sample below surface and for levelling, 4) adding 0.2 m (in one case
0.5 m) for river-dune basal peat data and 0.1 m for "ordinary" basal
peat data (assumed maximum water depth, see below), and 5) connecting
the lowest/youngest error boxes thus obtained.   Step 4 concerns the
indicative value of basal peat data and requires a brief explanation.
        Sea-level related basal peat growth is known to occur up to many
meters above MSL and even above MHW (Bennema, 1954; Jelgersma, 1961;
Van de Plassche, 1982).   The qualitative assessment of the bp-MSL
curve given above is based on the assumption that in a humid climate
initial peat growth on a pre-existing surface cannot occur below MSL.
However, while this assumption can be considered valid for basal peat
growth on a gently inclined surface (like most of the Pleistocene
subsurface in the western and northern Netherlands), there is no
reason why organic accumulation on the slope of a submerging river
dune in a deltaic plain should not be possible down to ca. 1 m below
MSL.   This can be expected to happen, for instance, when the local
tidal range approaches zero and water depth around the dune is between
1 and 2 m.   Assuming an initial situation in which the dune is not
surrounded by peat-forming vegetation, this can rapidly establish
itself from MSL to ca. 1 m below.   However, once reeds, rushes,
sedges, ferns and other plants have colonized the dune slope below
MSL, the upper limit of basal peat growth will, other factors
remaining constant, be maintained at a level near or slightly below
MSL during the subsequent rise of sea level.
        Thus, basal peat time-depth data can, in principle, only be used
to determine an upper limit for MSL rise.   Here it is assumed that if
a dune slope is continuously covered by peat over a vertical range of
more than 0.8 m, the base of the peat at and above that limit has
accumulated in a maximum water depth of 0.2 m, i.e. MSL was at most
0.2 m higher.   For other basal peat data a maximum water depth of 0.1
m is accepted, i.e., MSL was at most 0.1 m higher (but may have been
much lower).

4.3   Lower Limit for MSL

An error band for the lower limit of MSL rise (band B, Fig. 9) has
been established from various kinds of radiocarbon or archaeologically
dated indicators (other than base of basal peat data; see Louwe
Kooijmans, 1976 and Van de Plassche, 1982): the top of the basal peat,
the surface of a salt-marsh deposit, the height of a tidal creek levee
far inland and others.   The principle has been to first derive an

extreme lower limit, below which MSL cannot possibly have occurred
(curve C, Fig. 9), and then, data permitting, to establish a more
probable lower limit for MSL. For example, the second oldest index
point (4210-4050 cal BC) concerns the top of a tidal creek levee many
tens of kilometers inland on which human occupation has occurred. An
extreme value of -6.20 m Dutch Ordnance Datum (NAP)(derived from -5.85
± 0.35 m NAP) for the lower limit of MSL is obtained by 1) not
correcting for compaction of the sediments below the levee deposit, 2)
assuming a large local tidal range (unlikely far inland), and 3)
assuming a high level of sedimentation above local MHW. A more
probable value of -5.85 m NAP is obtained by compensating for some
compaction and assuming a slightly smaller local tidal range. The
error band is derived by connecting the more likely lower limits for
MSL thus obtained. Sharper lower limits might be established, but
only at the cost of decreasing reliability.

4.4   Conclusions

        1.   Basal peat data by themselves only give an upper limit for
MSL rise. The bp-MSL curve should therefore be replaced by an error
envelope (band A, Fig. 9) within or below which the true MSL rise
occurred.

        2.   It is apparent from Fig. 9, using available basal peat and
other back barrier sea-level data, that the MSL rise in the western
and northern Netherlands between 5000 and 500 cal BC can be
established with a minimum accuracy ranging from ± 0.75 m to ± 0.30 m
and with a maximum accuracy varying from ± 0.25 to less than ± 0.10
m.

4.5   Accuracy of the cb-MSL Curve (Fig. 10)

The accuracy of the cb-MSL curve depends on that of the MHW and MLW
curves. Due to the larger data set, the greater reliability of the
indicative features, and smaller vertical range among and between
them, the MHW curve clearly can be considered more reliable than the
MLW curve. Roep and Beets (1988) estimate the respective accuracies
at a few decimeters (MHW curve) and in the order of 0.5 m (MLW curve).
These values are principally based on visual inspection of the data
and of the lines connecting similar sedimentary features ("lithologic
trendlines"; see Roep and Beets 1988), rather than on systematic
evaluation of error sources involved. No such assessment is attempted
here; it suffices to mention a number of aspects and difficulties that
will have to be taken into consideration (see also Roep and Beets
1988).
        1. The cb-MSL curve is based on only 19 random samples from a
complex barrier system, unevenly distributed in space (an area of over
80 km^2) and time (large data gaps).
        2. The quality of the exposures (building pits and pipe-line
trenches), in terms of depth (completeness of the sequence), extent
(lateral variation in level of sedimentary features), detail visible

(reliability of the observations) and the time available for their study differed considerably from case to case.

3. Another factor that renders the data set more inhomogeneous, and therefore more difficult to evaluate than that of the bp-MSL curve, is that the data derive from two main types of progradational sequence (Beets et al., 1981); one representing open conditions (relatively steep nearshore gradients, slow progradation) and the other a more wave-protected situation (lower coastal gradients, rapid progradation). The first type, although probably formed in a setting comparable to that at present, is more difficult to interpret in terms of former sea-level stands than the second type, for which, however, no modern equivalent exists along the Dutch coast.

4. Where coastal barrier sands overlie clayey deposits, the possibility of sediment compaction should be considered.

5. The top of a coastal barrier sequence is younger than the base, the age difference being a function of the rate of progradation. In many cases the sequence has been dated by means of only one sample of shell doublets collected from the top or base of it. Specific rates of progradation, however, may be difficult to establish with confidence.

6. Because of wave asymmetry, HTL may be as much as 0.15 m above or below MSL.

5.   DISCUSSION AND CONCLUSION

Figure 11 compares the error bands for the upper and lower limit of the MSL rise, derived from basal peat and other non coastal barrier data, with the coastal barrier MHW, MSL (HTL) and MHW curves of Roep and Beets (1988). The question we set out to investigate, namely whether the good agreement between the bp- and cb-MSL curves is real or apparent, has in fact become meaningless because the bp-MSL curve has been replaced by a narrow error envelope for the upper limit of the MSL rise (band A, Figs. 9 and 11). Moreover, the error margins of the cb-MSL curve remain to be established. Nevertheless, it may be noted at this stage that, except for the oldest part of it, the cb-MSL curve neatly falls within the constraints for the rise of MSL as derived from available basal peat and other back barrier data. For the position of the original bp-MSL curve relative to the cb-MSL curve, see Figure 1.

Furthermore, it can be argued that the lower upper limit for MSL rise is much more probable than the most upper limit, for if the latter is assumed to be the true rise of MSL, the vertical distance to highest burrow levels and lowest base of dry eolian sand, as measured in the coastal barrier deposits, yields a coastal tidal range that is improbably small. This observation narrows down the constraints for the MSL rise to such an extent that it can be concluded that the original bp-MSL curve very probably is, as far as the trend is concerned, a reliable approximation of the MSL rise and that the good agreement with the cb-MSL curve is, for the larger part, likely to be real.

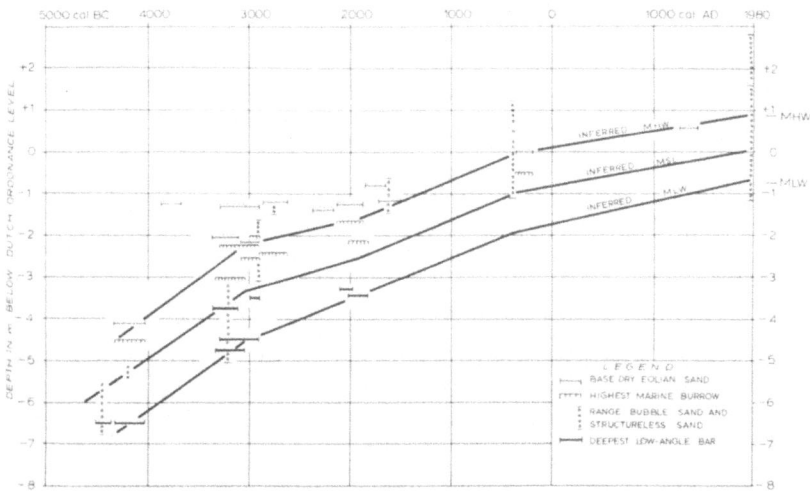

Figure 10.   MHW, MLW and MSL (HTL) curves based on coastal barrier
data (from Roep and Beets, 1988).

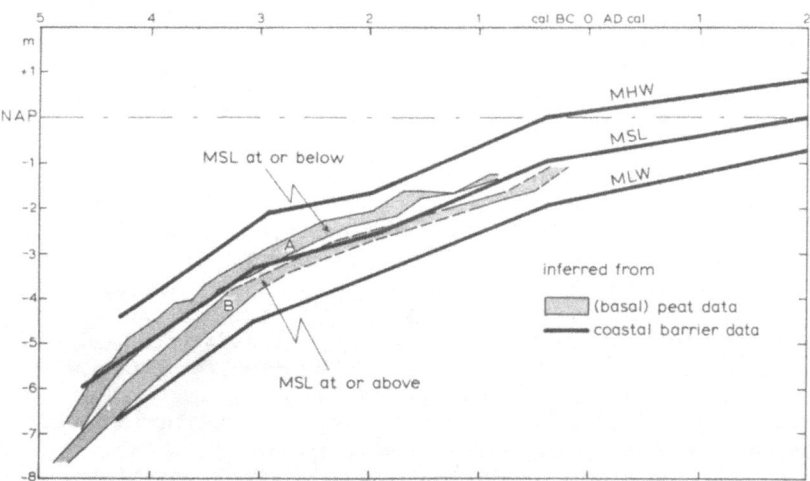

Figure 11.   Comparison of the coastal barrier MHW, MSL and MLW curves
(Roep and Beets, 1988) with the newly derived error bands for the
upper and lower limit of the MSL rise (Fig. 10) based on basal peat
and other back barrier sea-level data.

6. CURRENT AND PLANNED ACTIVITIES

New, unpublished sea-level data have been, and remain to be collected in the past and coming 3 to 5 years to fill in parts of the data gaps still existing. For instance, the trends of the MHW, MSL and MLW rise prior to 5300 yBP and since 2500 yBP need to be established with greater accuracy, and the question whether the SW Netherlands underwent or is undergoing less crustal subsidence than the western and northern parts of the country is still open.

As a contribution to IGCP Project 200, the first author has studied and sampled in detail the organic deposits surrounding the most westerly known, submerged late glacial river dune in the Rhine/Meuse delta. Located ca. 12 km west of Rotterdam, the 4 m of peat, mainly composed of reed, sedge and fern remains, show a regular alternation of at least 12 clayey and non-clayey, sometimes oxidized beds. The 20 samples that have been collected from the peat/dune-sand interface every 0.3 m between -8 and -4 m NAP are expected to yield valuable information on fluctuations in the rate of the local water-level rise between about 6300 and 4800 yBP. However, data older than ca. 5700 yBP will contain a river-gradient effect, which will complicate interpretation of their time-depth position in terms of MSL rise. To facilitate this interpretation, additional basal peat samples have to be collected at sites where a river- and/or groundwater-gradient effect is small or absent (e.g. in NE Friesland).

Within the frame of the so-called "Coastal Genesis" project, a joint, multi-disciplinary enterprise of the Rijkswaterstaat (Ministry of Transport and Public Works), the Netherlands Geological Survey (GSN), Delft Hydraulics, and several university institutes, aimed at the study of large-scale morphological behaviour of the Dutch coastline, Van der Valk (in close cooperation with the GSN and Roep) is currently carrying out a study of the development of the coastal barrier complex south of Haarlem (Van der Valk 1985). Sedimentological analysis of lacquer peels of barrier and shallow marine deposits is likely to yield additional information on former paleo-tide levels, particularly for the period 3500-2300 yBP.

Very few sea-level data are available for the last 2500 years. Participating in a regional sea-level project sponsored by the Commission for the European Communities, the GSN is making an effort to obtain data for this period, primarily from the Wadden Sea Islands.

Finally, new sea-level data remain to be collected in the Province of Zeeland in order to establish whether or not tectonic subsidence was or is smaller than further to the north. As mentioned above, available sea-level data suggest the possibility of slower crustal subsidence (Fig. 3). Jelgersma (1961) pointed out, however, that the data may also indicate the effect of a larger tidal range. In addition, Van de Plassche (1982) showed, in analogy to the situation in the mid-western Netherlands, that a higher time-depth position of the Zeeland data may be attributed, in part, to a groundwater-gradient effect.

7.   ACKNOWLEDGEMENTS

We thank Dr. S. Jelgersma for critical comments on the first draft of this paper. Mr. Henri Sion prepared the figures.

REFERENCES

Beets, D.J., Th.B. Roep and J. de Jong, 1981. Sedimentary sequences of the sub-recent North Sea coast of the western Netherlands near Alkmaar. In: S.-D. Nio et al. (eds.), IAS-Special Publication 5, 133-145.
Bennema, J., 1954. Bodem- en zeespiegelbewegingen inhet Nederlandse kustgebied. Thesis, Wageningen. Also in: Boor en Spade 7, 1-96.
Ente, P.J., 1976. The geology of the northern part of Flevoland in relation to the human occupation in the Atlantic time. Swifterband contribution 2 - Helinium 16, 15-35.
Griede, J.W., 1978. Het ontstaan van Frieslands Noordhoek. Thesis, Amsterdam, 186 pp.
Jelgersma, S., 1961. Holocene sea-level changes in the Netherlands. Thesis, Leiden. Also in: Meded. Geol. Stichting, C-IV (7), 100pp.
Jelgersma, S., 1966. Sea-level changes during the last 10.000 years. Proc. Int. Symp. World Climate from 8000-0 BC, Roy. Meteor. Soc. London, 54-71.
Jelgersma, S., 1977. Zeespiegelbeweging en bodemdaling. In: C.J. van Staalduinen, (ed.).: Geologisch onderzoek van het Nederlandse Waddengebied. Rijks Geol. Dienst, Haarlem, 72-74.
Jelgersma, S., 1979. Sea-level changes in the North Sea basin. In: E. Oele et al. (eds.): The Quaternary history of the North Sea, Acta Univ. Ups. Symp. Univ. Ups. Annum Quingentesium Celebrantis 2, 233-248.
Jelgersma, S., 1980. Late Cenozoic sea-level changes in The Netherlands and the adjacent North Sea basin. In: N.-A. Morner (ed.), Earth rheology, isostasy and eustasy, John Wiley and Sons (London), 435-447.
Jelgersma, S., Th.B. Roep and D.J. Beets, 1975. New data on sea-level changes in the Netherlands. Guidebook INQUA Shoreline meeting Belgium, Netherlands, Germany.
Louwe Kooijmans, L.P., 1974. The Rhine/Meuse Delta; four studies on its prehistoric occupation and Holocene geology. Thesis, Leiden, 421 p.
Louwe Kooijmans, L.P., 1976. Prahistorische Besiedlung im Rhein-Maas-Deltagebiet und die Bestimmung ehemaliger Wasserhohen. Probleme der Kustenforschung im sudlichen Nordseegebiet 11, 119-143.
Pearson, G.W., J.R. Pilcher, M.G.L. Baillie, D.M. Corbett and F. Qua, 1986. High-precision ^{14}C measurements of Irish oaks to show the natural ^{14}C variations from AD 1840 to 5210 BC. In: M. Stuiver & R.S. Kra (eds.), Calibration Issue, Radiocarbon 28, 2 B, 911-934.

56

Pons, L.J., S. Jelgersma, A.J. Wiggers and J.D. de Jong, 1963. Evolution of the Netherlands coastal area during the Holocene. Verh. Kon. Ned. Geol. Mijnb. Gen., Geol. Serie 21-2. 197-208.

Roeleveld, W., 1974. The Groningen coastal area. A study in Holocene geology and lowland physical geography. Thesis, Amsterdam. Also in: Bericht. Rijksdienst Oudheidk. Bodemonderz., 20-21, 1970-'71 and 24, 1974, Supplement.

Roep, Th.B., 1986. Sea-level markers in coastal barrier sands: examples from the North Sea coast. In: O. van de Plassche (ed.): Sea-level research: a manual for the collection and evaluation of data, Geo Books (Norwich), 97-128.

Roep, Th.B., and D.J. Beets, 1988. Sea-level rise and paleotidal levels from sedimentary structures in the coastal barrier in the western Netherlands since 5600 B.P. Geologie en Mijnbouw, 67,1, 53-60.

Roep, Th.B., D.J. Beets and G.H.J. Ruegg, 1975. Wave-built structures in subrecent beach barriers of The Netherlands. Proc. IXth Int. Congr. of Sedimentology, IAS, Nice 1975, 6, 141-145.

Roep, Th.B., D.J. Beets and J. de Jong, 1979. Het zeegat tussen Alkmaar en Bergen van ca 1900 tot 1300 jaar voor Chr. Alkmaarse Historische Reeks III, 9-35.

Roep, Th.B., O. van de Plassche, L. van der Valk and G.H.J. Ruegg, 1984. Sedimentologie van de strandwalafzettingen onder 's-Gravenhage en Rijswijk. In: H.W.J. van Amerom (ed.): De bodem van 's-Gravenhage, Meded. Rijks Geol. Dienst 37-1, 63-95.

Shennan, I., 1987. Holocene sea-level changes in the North Sea region. In: M.J. Tooley & I. Shennan (eds.), Sea-level changes, Oxford: Blackwell, 109-151.

Van de Plassche, O., 1979. Sea-level research in the Province of South Holland, Netherlands. In: K. Suguio et al. (eds.): Proc. 1978 Int. Symp. Coastal Evolution in the Quaternary (Sao Paolo, Brazil), 534-551.

Van de Plassche, O., 1981. Sea-level, groundwater, and basal peat growth - a reassessment of data from The Netherlands. In: A.J. van Loon (ed.): Quaternary geology: a farewell to A.J. Wiggers, Geologie en Mijnbouw 60, 401-408.

Van de Plassche, O., 1982. Sea-level change and water-level movements in The Netherlands during the Holocene. Thesis, Amsterdam. Also in: Meded. Rijks Geol. Dienst 36-1, 1-93.

Van Straaten, L.M.J.U., 1954. Radiocarbon dates and changes of sea level at Velzen (Netherlands). Geologie en Mijnbouw 16, 247-253.

Van der Valk, L., 1985. Coastal barrier deposits in The Hague, South Holland, the Netherlands. In: Abstracts Symp. Modern and ancient clastic tidal deposits, Utrecht 1985, 149.

Van der Valk, L., W.E. Westerhoff and J. de Jong, 1985. Mid-Holocene wave dominated clastic tidal deposits at Rijswijk (South Holland) in the western parts of the Netherlands; a case study. In: Abstracts Symp. Modern and ancient clastic tidal deposits, Utrecht 1985, 145-148.

Vogel, J.C. and H.T. Waterbolk, 1963. Groningen radiocarbon dates 4. Radiocarbon 5, 163-202.

RESPONSE OF SANDY BEACHES TO SEA-LEVEL RISE

Stephen P. Leatherman
Laboratory for Coastal Research &
Department of Geography
University of Maryland
College Park, MD 20742

ABSTRACT. The response of sandy shorelines to sea-level rise is
erosion unless offset by ample sediment supply. Geomorphic indicators
of shore retreat include beach, dune, and cliff erosion as well as
backbarrier peat outcrops on the barrier beachface. Quantitative
analysis of historical maps and aerial photographs also indicate pervasive erosion along the U.S. clastic coastlines. The principal
approaches to determine shore retreat in response to sea-level rise
are trend analysis, Bruun rule, and sediment budget model. Societal
responses fall into three general categories: fortify the shoreline,
nourish the beach, or retreat from the coast.

INTRODUCTION

The underlying cause of shoreline displacement (transgressions and
regressions) is water level changes. Over geologic time, sea levels
have fluctuated hundreds of meters. While most IGCP studies have
involved the use of ancient shores to infer past sea levels, this
approach is to investigate the effects of past and hence future
increases in sea level on shoreline change.

The relative change in sea level at a given location is the sum
of eustatic effects, regional influences, and local phenomena. There
are essentially six long-term causes of relative sea-level rise:
eustatic component, crustal and seismic subsidence, natural and human-
induced compaction, and variations due to climatic fluctuations. The
eustatic component has been much discussed, and the term is perhaps a
misnomer by definition (Morner, 1976). Here it is used to refer to
glacio-eustasy due to melting of land-based glacier ice and steric
expansion of near-surface ocean water due to global warming. Climatic
effects, such as El Nino-Southern Oscillation, are discussed elsewhere
(Barnett, 1983).

Of the four subsidence processes, only anthropogenic subsidence,
due to withdrawal of subsurface fluids, can be reversed or at least
partially mitigated by recharge or other means. Crustal subsidence of
the land surface due to neotectonics is a contemporary structural

D. B. Scott et al. (eds.), Late Quaternary Sea-Level Correlation and Applications, 57–69.
© 1989 by Kluwer Academic Publishers.

downwarping of the earth's crust due to a variety of geologic processes (Newman et al, 1983). Of particular interest along the U.S. East Coast is the subsidence of the former marginal forebulge area, responding to the southerly advance to New Jersey and then northwardly retreat of the glaciers during the Wisconsinan glaciation.

Neotectonics maps can be used on a regional scale to delineate areas of uplift or down-warping due to modern crustal processes. For instance, Newman et al (1980) have compiled a large data set on radio-carbon dates of sea-level adjusted peat deposits to derive a computer-plotted isobase map for the past 1,500 years (Figure 1). This map can be used to explain the apparent anomaly of rapid relative sea-level rise recorded for the Atlantic City, N.J., tide gauge during the last century. When the assumed eustatic component is subtracted from the trend, it is clear that two-thirds of the relative rise can be attributed to site-specific conditions (local compaction of uncon-solidated coastal sediments) plus the regional framework (neotectonic activity).

Seismic subsidence of the land surface occurs suddenly due to earthquake activity, which is of particular importance along the Pacific coast rim. Autosubsidence represents the compaction of soft, unconsolidated sediments, especially mud or peat, at or beneath the land surface.

Considering the variety of causes of relative sea-level rise, it is not unexpected that worldwide records indicate a large range in water level changes during the past hundred years. Of the various observational techniques, tide gauges provide the most detailed record, but the data require careful analysis. In order to delineate the eustatic component from the record of relative sea-level rise, Very Long Baseline Interferometry (VLBI) in combination with differ-ential Global Positioning Systems (GPS) is being employed (Carter et al, 1986). These analyses provide decade to century-long trends, with geologic indicators extending the record to millenia. This approach is particularly valuable in demonstrating the natural, long-term response of a coastal sector as compared with the shorter-term time series data provided by geodetic leveling and tide gauges.

GEOMORPHIC INDICATORS OF EROSION

The geologic record is replete with examples of major transgressions and regressions. The U.S. Atlantic and Gulf coastal plains are the sedimentary record of these processes operating over the last ten million years as the continental margins have slowly subsided. Sea levels have fluctuated by several hundred meters during this time period. Five distinct transgressive coastal systems have been identi-fied on the U.S. Delmarva peninsula using geomorphic and subsurface data. Each was produced during interglacial high sea levels and range in age from over one million years to 60,000 years B.P. (Demarest and Leatherman, 1985). Sedimentological and historical evidence for four

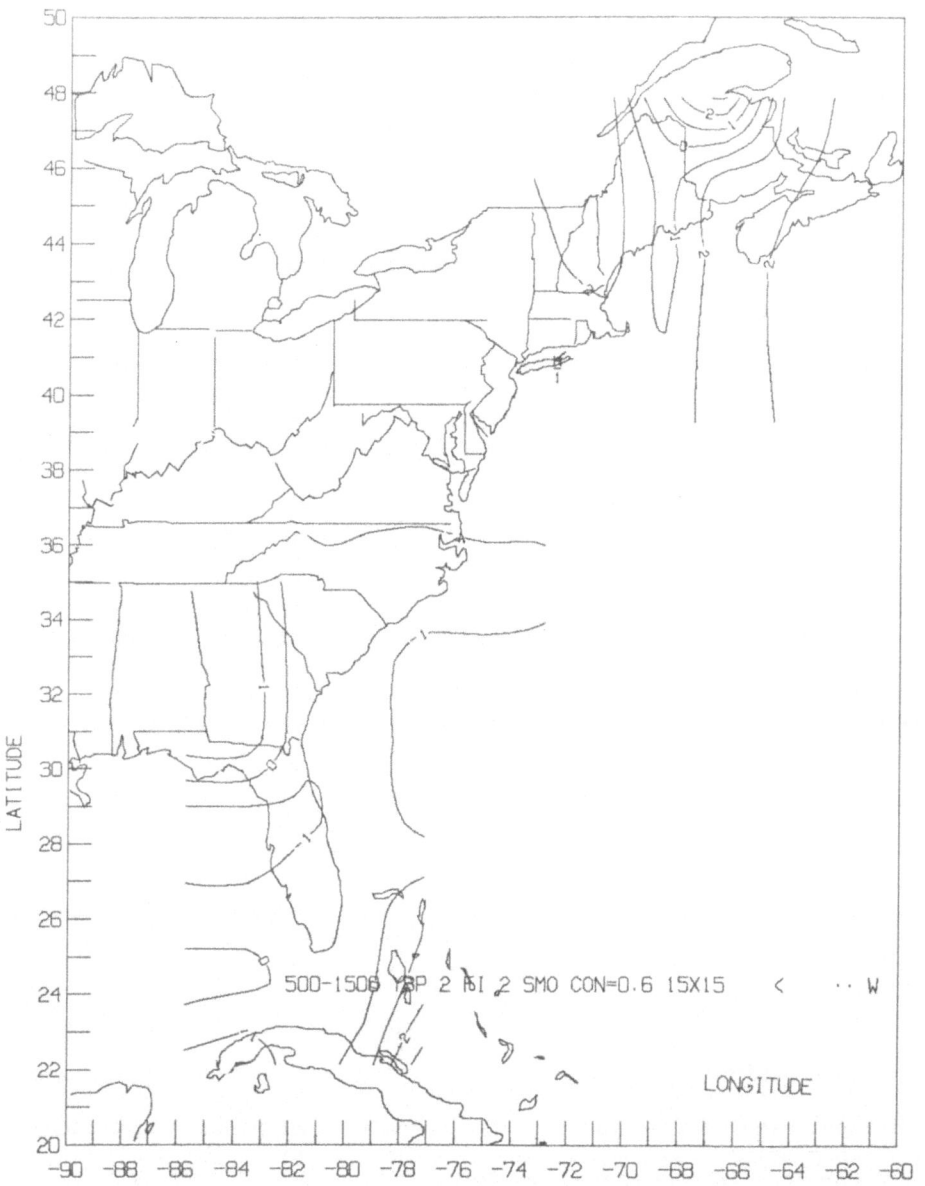

Figure 1. Isobase map of the U.S. Atlantic Coast (1,000 years
B.P.). During the last few thousand years, the trend of relative mean
sea level along the U.S. East Coast has been rising in the Bahamas,
Cuba and Hudson Bay, Canada and down in Nova Scotia and the Mid-
Atlantic Coast. (from Newman, unpublished data)

minor transgressive phases or pulses during the last 2,000 years, with sea-level fluctuations of less than a meter, have been reported along the Friesland barrier islands in the Netherlands (Bakker, 1981). The modern transgressive pulse during the overall Holocene transgression presumably began in the late 18th century, a period marked by increased storm surge damage and coastal flooding.

Long periods of sedimentary accretion resulting in beach ridges have been arrested or the trend reversed during the past century. Teichert (1947) reported that beach ridge formation ceased slightly more than one hundred years ago, and the Western Australian coralline shore is now subject to erosion by the sea. In Nigeria, Pugh (1954) noted that earlier progradation had similarly given way to retrograda-tion on sandy shorelines. Bogue Banks along the Outer Banks of North Carolina is a barrier island composed of parallel sets of beach ridges, which have prograded seaward during the past 3-4 thousand years, but now the beaches are narrow and dunes are actively cut by waves during annual winter storms (Steele, 1980). Similar reversals in trend from long-term accretion to recession have been noted by many investigators working along sedimentary coasts in the U.S. (e.g., Tanner and Stapor, 1971) and worldwide (Davies, 1957).

Onset of the present transgressive pulse, attested to by marked beach and dune erosion, has varied geographically due to local dif-ferences in sand supply and wave energy. Information supplied by 73 correspondents from 39 coastal countries showed that less than 10 percent of the length of the world's sandy shorelines had prograded, more than 60 percent had retrograded, and the rest have been rela-tively stable or shown no consistent trend during the past century (Bird, 1976).

Other geologic indicators of shore retreat are wave-cut cliffs, which occur essentially worldwide. Exhumation of salt marsh peat on the beachface indicates upward and landward barrier migration. Most barrier island coasts have been retreating for at least the past few hundred years as clearly indicated by these backbarrier peat outcrops, exposed on the lower beach foreshore after severe storms. Peat out-crops have been reported from widely dispersed areas along the U.S. Atlantic and Gulf Coasts (e.g., Leatherman, 1979a, 1979b; Kraft, 1971).

HISTORICAL RECORDS

Historical records also indicate the prevalence of shore retreat along U.S. sandy beaches during at least the past century. The National Shoreline Study by the U.S. Army Corps of Engineers (1971) was the first overall appraisal of U.S. shore erosion problems. Of the 84,000 miles of United States ocean and Great Lakes shorelines, significant erosion occurs along 43 percent of the total, excluding Alaska.

More recently, May et. al (1983) have assembled the extant data for the U.S. coastline. Most of their information is based on his-torical aerial photography dating back to the late 1930s so that a maximum record of 50 years is available. The National Ocean Service of NOAA and University of Maryland researchers (Leatherman, 1983a) have made major advances in quantifying the historical rate of shore-line change along large portions of the U.S. continental coast. His-torical maps and charts (NOS "T" sheets), updated with aerial photo-graphs, extend the record to more than 100-150 years. These data, which include most of the U.S. Northeast and mid-Atlantic coasts as well as South Carolina and parts of California, indicate a general pattern of pervasive shore retreat during the past century.

TECHNIQUES OF PROJECTING SHORE RETREAT

There are several difference approaches that can be used to determine the amount of shore retreat as a function of sea-level rise. The simplest approach to apply is the "drowned valley" concept, whereby pre-existing, upland topography is used to project new shorelines. This is the preferred methodology for immobile substrates or where the wave climate is subdued.

Rising sea level is accompanied by a general retreat of the shoreline due to inundation and/or erosion, except where the effects are offset by ample sand supplies. Erosion is the physical removal of beach material, while inundation is the submergence of the otherwise unaltered shore. The response time is instantaneous with landward movement of the shoreline due to simple inundation of the land. While the approach is straightforward, submergence accounts for only a small portion of the net shore retreat along exposed, clastic coasts.

The principal approaches to projecting shore retreat that have been employed to date are largely based on the erosional potential of sea-level rise: (1) extrapolation of trend (Leatherman, 1985), (2) Bruun Rule (Hands, 1981), and (3) sediment budget model (Everts, 1985).

Historical Trend Analysis

Trend analysis involves the empirical determination of projecting new shorelines, wherein shore response is based on the historical trend with respect to local sea-level change during that time period. This procedure takes into account the inherent variability in shoreline response based on differing coastal processes, sedimentary environments, and coastline exposures.

The method of projecting shoreline movement due to sea-level rise involves (1) quantification of the historical record of shoreline movement (using accurate maps, charts, and vertical aerial photo-graphs), (2) determination of a historical shore retreat to sea-level rise relationship for different shorelines, and (3) establishment of an hypothesis or rule of thumb on the basis of which to project further sea-level rise - shoreline movement relationships (Leatherman, 1985). In this case, it is assumed that the amount of retreat determined from the historical record is directly correlated with the

rise rate of sea level, assuming lag effects in shoreline responses
are small compared to overall extrapolation accuracy.

Tide gauge records (Hicks et al, 1983) document the local
(eustatic plus isostatic effects, such as subsidence) rate of sea-
level change over the period of record. For example, the Galveston,
Texas tide gauge records indicate that sea level is rising at an
adjusted rate of approximately 30 cm/century. Historical shoreline
data of mappable quality are available from 1850 to present for the
Galveston Bay area. National Ocean Service (NOS, then called U.S.
Coast Survey and U.S. Coast and Geodetic Survey) shoreline manuscripts
were used for shoreline comparisons. The NOS "T" sheets were made
from field surveys and are the most accurate maps of the shoreline
presently available. The Metric Mapping procedure (Leatherman, 1983a)
was used for data compilation and computer mapping. The results of
this analysis are presented elsewhere (Leatherman, 1983b).

This type of analysis can be undertaken for any coastal plain
shoreline. The easily-eroded, unconsolidated sediments and gently-
sloping, low-lying topography make the projections straightforward,
except where modified by coastal engineering structures. The underlying assumption of this analysis is that shorelines will respond in
similar ways in the future to sea-level rise if all other parameters
remain essentially constant.

Total shoreline adjustment to sea-level rise at the particular
scenario year is assumed by this approach. Obviously, there will be
some lag in shoreline response to higher water levels. Each area will
have a different lag time in shoreline response depending upon storm
frequency. Simple extrapolation of historical trends is deemed a
reliable, first cut approach for forecasting shoreline changes.

Bruun Rule

Bruun (1962) was the first to evaluate quantitatively the role of
rising sea level in shore erosion. His formulation is based on the
concept of an equilibrium profile, which he defined as a statistical
average profile which maintains its form, apart from small fluctuations, at a particular water level. Use of the term equilibrium in
this context is consistent with the recognition of seasonal, storm, or
other temporal profile fluctuations.

The Bruun Rule provides for a profile of equilibrium in that the
volume of material removed during shoreline retreat is transferred
onto the adjacent shoreface/inner shelf, thus maintaining the original
bottom profile and nearshore shallow water conditions (only further
inland). Hence the formulation represents an on/offshore sediment
balancing between eroded and deposited material without consideration
of longshore losses. With an incremental rise in sea level, it is
clear that additional sand must be added to the below-water portion of
the beach profile; negating longshore variations, this sand must be
derived from beach erosion.

Bruun's concept is straightforward and intuitively appealing, but
it is difficult to confirm or quantify without precise bathymetric

surveys and integration of complex nearshore profiles over a long
period of time. Also, definition of the active profile boundaries in
the seaward direction necessitates the selection of a pinch-out depth
of significant sediment motion, a rather vague concept in an oceanic
wave environment. The problem is confounded when an attempt is made
to quantify a relatively confined zone of erosion (e.g., the narrow
beach/dune zone) with a broad zone (the shoreface/inner shelf) over
which eroding sediment can be thinly spread.

Hallermeier (1981) has shown that the so-called "depth of
closure" or pinch-out depth varies considerably around the United
States shorelines. The 10 meter definition of wave base is adequate
as a first approximation (Dietz, 1963), but this average depth cannot
be applied to specific areas to obtain quantitative results. Definition of the limiting water depth to the active beach profile is
required for prediction of shore erosion caused by sea-level rise.
Hallermeier (1981) demonstrated that this depth value depended on some
specific nearshore wave height statistic; typical values for the U.S.
continental coasts range from 4-8 meters.

Hands (1976) introduced a simplification of the Bruun (1962)
method for application to the Great Lakes, wherein $x=zX(R_a) \div Z$, where
x = shoreline change, z = change in water level, X = average, representative width of adjustment in the profile, Z = height of responding
profile or vertical relief of active beach, and R_a = overfill ratio to
account for loss of suspended load to the offshore from the eroded
material. This approach assumes that the profile will return to a
specific shape after being influenced by sea-level rise (in accordance
with Bruun's (1962) equilibrium profile). The profile closure depth,
beyond which the bottom does not respond to surface changes, depends
on local wave climate.

The Great Lakes served as a natural laboratory to document the
effects of rising water levels on shore position. Hands (1981) found
that the Bruun Rule was well satisfied as quantified by field surveys
of beach and nearshore profiles during rising water levels in Lake
Michigan. The volume of sand eroded from the beach nearly matched
offshore deposition (Figure 2), providing the first actual field veri-
fication of this hypothesis.

Profile retreat was found to lag behind lake-level rise; response
rates are tied to storm intensity and frequency. Rising water levels
establish a potential for erosion, but realization of the potential
requires sediment redistribution, i.e., work which depends on energy
being available. Perhaps a constructive way of viewing the allied
roles of sea-level position and sea energy is to consider that sea
level sets the stage for profile adjustments by coastal storms. Long-
term sea-level rise places the beach/nearshore profile out of equilibrium, and sporadic storms accomplish the geologic work in a quantum
fashion. Major storms are required to stir the bottom sands at great
depths offshore and hence fully adjust the profile to the existing
water level position. Therefore, the underlying assumption is that
beach equilibrium will be the result of water level position in a
particular wave climate setting.

Sediment Budget Approach

The Bruun Rule is essentially sediment balancing in a two-dimensional sense, which should be theoretically reached assuming no gradients in the longshore transport system or offshore leakage along a long, otherwise uninterrupted sandy shoreline. However, littoral drift rates are an order of magnitude larger than on/offshore transport for most coastal areas (Dean, 1976). Other losses to the nearshore sand-sharing system, such as inlet shoals, must be considered, therefore formulation of a sediment budget is required to quantify the three-dimensional changes.

The sediment budget approach essentially represents a conservation of sand based upon an understanding of sources, sinks and fluxes. While fundamentally straightforward in approach, application is hampered by data requirements (hard numbers on an annual basis for littoral drift, inlet losses, and offshore leakage). Quantification of nearshore and shoreface profile changes is particularly problematic since small vertical changes over such broad areas can represent huge volumetric amounts, and field measurements are probably the least reliable in this zone.

Everts (1985) developed a sediment budget analysis, which considers losses and gains of sand within a bounded coastal reach (e.g., control volume, Figure 3). He assumed, as did Bruun (1962), that the beach profile will remain in equilibrium, moving landward and upward in time and space with sea-level rise. The sediment balance required to maintain an equilibrium profile, which does not change shape, proceeds from a consideration of net losses or gains to the control volume. Sea-level rise must be considered, and the effects averaged, over a relatively long time period (at least several decades for the open coast). Also, the sediment budget must be considered regional in scope; a good approach is to define a coastal compartment between natural or human-induced littoral barriers, such as tidal inlets. The relative sea-level rise data are derived from the nearest reliable tide gauge over the period of time considered. A factor must be entered into the sediment budget to account for the quantity of fine-grained material that subcrops on the shoreface and is contained in the barrier. This material will be quarried by the wave beveler with barrier transgression (Leatherman, 1983c), and the percentage of coarse sediment available is a prime consideration when a barrier retreats landward by feeding upon itself.

Everts (1985) estimated losses to the shoreface (offshore leakage in excess of equilibrium profile) by considering nearshore shoals as sand losses to the landward retreating barrier. Field and Duane (1976) have shown that these linear sand ridges, believed to have initiated on the beach shoreface, are dynamic, rather than relic features. With continued sea-level rise and concommitant landward barrier retreat, these large sand bodies are left behind, albeit reworked surficially, and eventually become detached from the near-shore, sand-sharing system, thus representing a net loss of sediment to the barrier system. The amount lost in this manner was estimated

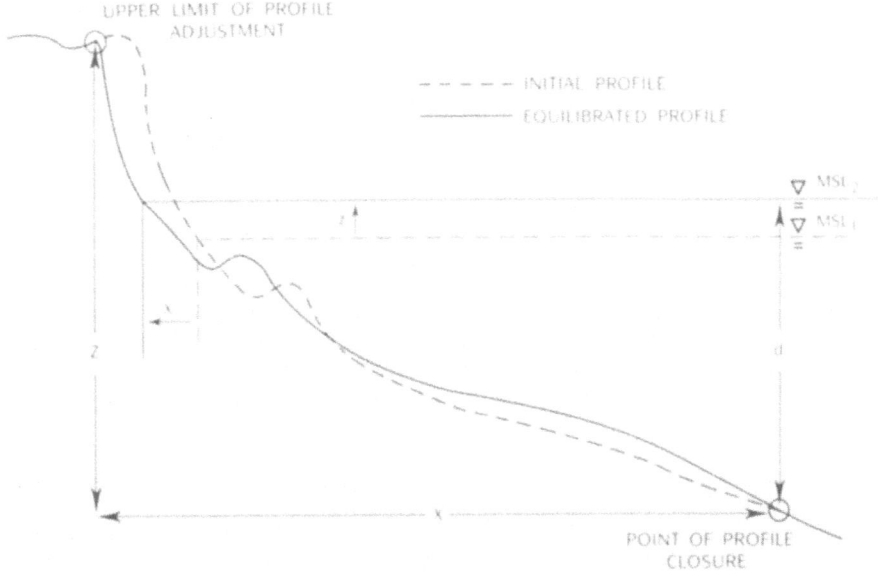

Figure 2. Profile adjustment to rising water levels, resulting in beach erosion and offshore deposition (from Hands, 1976).

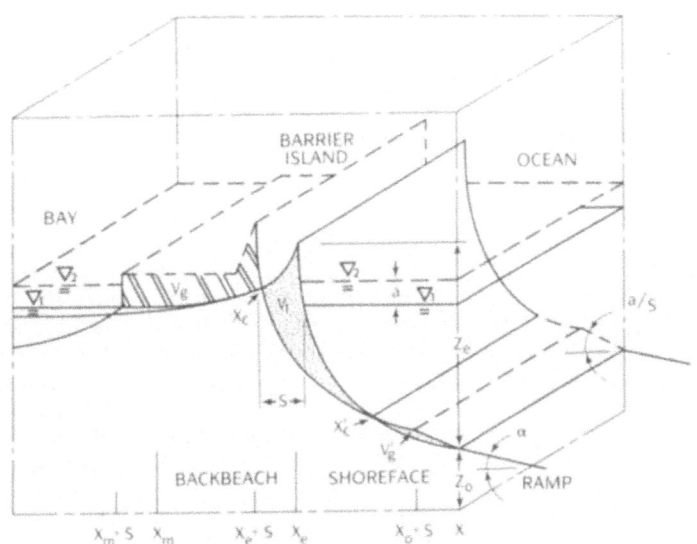

Figure 3. Control volume approach to computing beach erosion (from Everts, 1985).

by volumetric calculations of the sand contained in closed contours
above average sea bed elevations obtained from recent bathymetric
charts. Other losses or gains, such as overwash and aeolian transport, beach nourishment and sand mining, can be reasonably well
determined from past geological studies and public records.

DISCUSSION AND CONCLUSIONS

It is apparent that the erosion problem along the world's sandy coasts
is a major concern. Climatologists predict that the greenhouse effect
will accelerate the rate of sea-level rise by at least two to five
times compared to the historical record. This will likely increase
the rate of beach erosion by similar amounts in a few decades. The
magnitude of these effects will depend upon the erodability or
inundation potential of a particular shore reach and the current and
projected population density and development activities.

For sandy coasts, the possible societal responses fall into three
general categories: (1) fortify the shoreline, (2) nourish the beach,
or (3) retreat from the water's edge. Each of these options needs to
be evaluated carefully within the context of the coastal geomorphic
setting (hence the cost of various shoreline engineering practices)
and the benefit (value of the property to be protected or left to
erode away). The choices will not be easy, and decisions must be made
on the basis of site-specificity evaluated at a particular point in
time.

Geologic and geomorphic analyses will play an important role in
the decision to defend or abandon the coast. The quantitative models
for predicting the amount of shore erosion with a particular rate of
sea-level rise must be improved and thoroughly tested with extensive
field data and historical map comparisons. This is an exciting time
to be undertaking such research since the stakes are so high and world
attention is focussing on the issues of climatic change and associated
changes in sea level.

ACKNOWLEDGEMENTS

This research was supported by the U.S. Environmental Protection
Agency, Washington, D.C. Some of this material appeared in the
recently published book, Responding to Changes in Sea Level:
Engineering Implications, 1987, National Academy of Sciences Press.
The author was a member of this committee and a principal author of
this book. This paper was delivered at the 1987 IGCP meeting in
Halifax, Canada as a keynote presentation, and travel support through
the NATO Advanced Studies Institute is gratefully acknowledged.

REFERENCES

Bakker, J.P., 1981. "Transgression phases and the frequency of storm floods in Netherlands in recorded times," in Overwash Processes, S.P. Leatherman, ed., Benchmark Papers in Geology, V. 58, Hutchinson Ross Publ. Co., p. 51-56.

Barnett, T.P., 1983. "Recent changes in sea level and their possible causes," Climate change, V. 5: 15-38.

Bird, E.C.F., 1976. "Shoreline changes during the past century," Proc. of 23rd International Geog. Congress, Moscow, 54 pp.

Bruun, P., 1962. "Sea-level rise as a cause of shore erosion," Amer. Soc. Civil Engr. Proc., Jour. Waterways & Harbors Division, V. 88: 117-130.

Carter, W.E., Robertson, D.S., Pyle, T.E., and Diamante, J., 1986. "The application of geodetic radio interferometric surveying to the monitoring of sea level," Geophys. J.R. astr. Soc., V. 87: 3-13.

Davies, J.L., 1957. "The importance of cut and fill in the development of beach sand ridges." Australian J. Science, V. 20: 105-111.

Dean, R.G., 1976. "Beach erosion: causes, processes, and remedial measures," CRC Critical Reviews in Environmental Control, p. 259-296.

Demarest, J.M. and S.P. Leatherman, 1985. "Mainland influence on coastal transgression: Delmarva peninsula," Marine Geology, V. 63: 19-33.

Dietz, R.S., 1963. "Wave-base, marine profiles of equilibrium, and wave-built terraces: a critical appraisal," Geol. Soc. Am. Bull., V. 74: 971-990.

Everts, C., 1985. "Effect of sea-level rise and net sand volume change on shoreline position at Ocean City, Maryland," U.S. Environmental Protection Agency Report, Washington, D.C., pp. 67-98.

Field, M.E. and D.B. Duane, 1976. "Post-Pleistocene history of the United States inner continental shelf: significance to origin of barrier islands," Geol. Soc. of Am. Bull., V. 87: 691-702.

Hallermeier, R.J., 1981. "A profile zonation for seasonal sand beaches from wave climate," Coastal Engineering, V. 4: 253-277.

Hands, E.B., 1976. "Observations of barred coastal profiles under the influence of rising water levels, eastern Lake Michigan, 1967-1971," U.S. Army Coastal Engr. Res. Ctr. Tech. Rpt. TR-76-1.

Hands, E.B., 1981. "Predicting adjustments in shore and offshore sand profiles on the Great Lakes," U.S. Army Coastal Engr. Res. Ctr. CETA 81-4., 25 p.

Hicks, S.D., H.A. Debaugh, Jr., and L.E. Hickman, Jr., 1983. "Sea level variations for the United States, 1855-1980," NOAA Report, Rockville, MD, 170 pp.

Kraft, J.C., 1971. "Sedimentary facies patterns and geologic history of a Holocene marine transgression," Geol. Soc. Amer. Bull., V. 82: 2131-2158.

Leatherman, S.P., 1979a. "Migration of Assateague Island, Maryland, by inlet and overwash processes," Geology, V. 7: 104-107.

Leatherman, S.P., ed., 1979b. Environmental Geologic Guide to Cape Cod National Seashore, SEPM-ES Special Publication, 249 p.

Leatherman, S.P., 1983a. "Shoreline mapping: a comparison of techniques," Shore and Beach, V. 51: 28-33.

Leatherman, S.P., 1983b. "Geomorphic effects of projected sea-level rise: a case study of Galveston Bay, Texas," Proceedings of Coastal Zone 83, ASCE, San Diego, CA, p. 2890-2901.

Leatherman, S.P., 1983c. "Barrier island evolution in response to sea-level rise: a discussion," J. of Sed. Petr., V. 53: 1026-1033.

Leatherman, S.P., 1985. "Geomorphic effects of accelerated sea-level rise on Ocean City, Maryland," U.S. Environmental Protection Agency, Washington, D.C., 34 pp.

May, S.K., R. Dolan and B.P. Hayden, 1983. "Erosion of U.S. shorelines," EOS Transactions, Am. Geophysical Union, V. 64: 551-552.

Morner, N.A., 1976. "Eustasy and geoid changes." Journal of Geology, V. 84: 123-151.

Newman, W.S., Marcus, L.F., Pardi, R.R., Paccione, J.A., and Tomacek, S.M., 1980. "Eustasy and deformation of the geoid: 1,000-6,000 radiocarbon years B.P.," in Earth Rheology, Isostasy and Eustasy, N.A. Morner, ed., John Wiley, New York, pp. 555-567.

Pugh, J.C., 1954. "A classification of the Nigerian coastline," J. West African Sci. Assoc., V. 1: 3-12.

Steele, G.A., 1980. "Stratigraphy and depositional history of Bogue Banks, North Carolina," M.S. thesis in geology, Duke University, Durham, N.C., 201 p.

Tanner, W.F. and F.W. Stapor, 1971. "Tabasco beach ridge plain: an eroding coast," Trans. Gulf Coast Assoc. Geol. Socs., V. 21: 231-232.

Teichert, C., 1947. "Contemporary eustatic rise in sea level?" Geogr. Journal, V. 109: 288-289.

U.S. Army Corps of Engineers, 1971. "National Shoreline Study," Washington, D.C. 59 p.

A LATE PLEISTOCENE LOW SEA-LEVEL STAND OF THE SOUTHEAST CANADIAN OFFSHORE

Gordon B. J. Fader
Atlantic Geoscience Centre
Geological Survey of Canada
Bedford Institute of Oceanography
P.O. Box 1006
Dartmouth, N. S.,
B2Y 4A2

ABSTRACT. A widespread, submarine, low sea-level stand is interpreted to occur at a present depth of 110-120m across the Scotian Shelf and is dated at 15 ka BP. Evidence in support of this position includes: 1) absence of fine-grained muddy sediments above and their abundance below 110m; 2) the distribution of well-sorted sand and rounded gravel clasts above the low sea-level position in contrast to the widespread occurrence of angular clasts below; 3) the occurrence of terraces cut into both bedrock and glacial sediments at 110-120m water depth; 4) the occurrence of unconformities on glacial sediments dated from 40-15 ka BP which exist as erosional remnants above the low sea-level position; 5) the distribution of continuous deposits of till below and its general absence above the low sea-level stand; 6) the relative distribution of relict versus modern iceberg furrows at the seabed above and below the low sea-level stand; and 7) the occurrence of a subaerially desiccated crust above 110m water depth and its absence below. The low sea-level position occurs at approximately the same depth across the outer continental shelf. This suggests that the glaciers had retreated from the shelf before the maximum lowering and that differential warping has not occurred since its formation, otherwise the low stand would be discontinuous. It also supports the idea that for the offshore area of the southeast Canadian continental shelf, glacial isostatic rebound was largely over by the time the low sea-level stand was formed.

INTRODUCTION

A widespread low sea-level position at a present depth of 110-120m has been identified along the Scotian Shelf and adjacent areas off southeast Canada. Through a regional sediment mapping program of the Geological Survey of Canada, this low level position has been determined on the basis of seabed morphology and stratigraphy along thousands of kilometers across the continental shelf and may represent one of the most extensively known low sea-level positions. The detailed sedimentological and acoustical supporting evidence for the low sea-level position was first documented by King (1967) for the central Scotian Shelf. This paper is an attempt to consolidate and review some of the earlier evidence on which the interpretations were based and to illustrate and discuss some of the most recently collected data.

Geological Survey of Canada Contribution No. 16388.

D. B. Scott et al. (eds.), Late Quaternary Sea-Level Correlation and Applications, 71–103.
© *1989 by Kluwer Academic Publishers.*

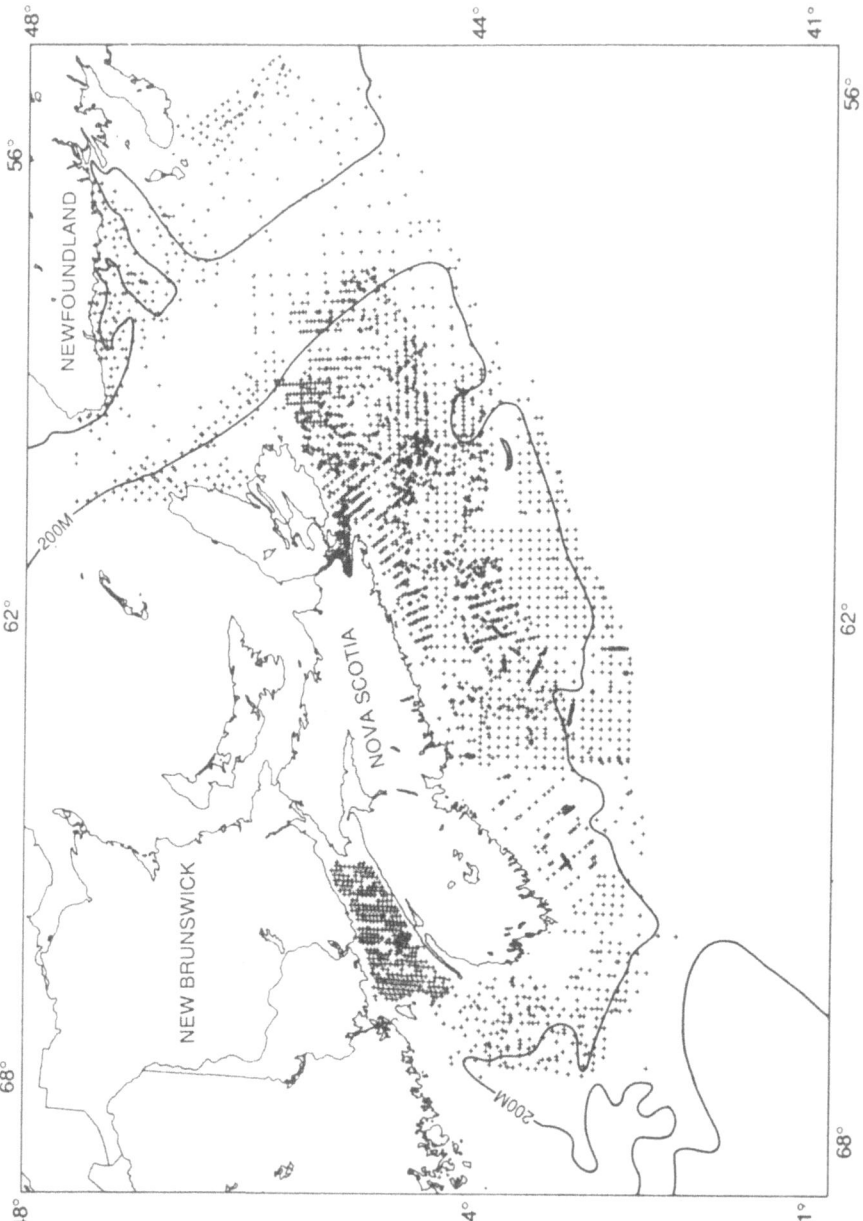

1. Distribution of samples from the Scotian Shelf and the western Grand Banks of Newfoundland collected for the compilation of the surficial sediment distribution maps.

PREVIOUS WORK

An understanding of the Late Pleistocene-Holocene sea-level history has important implications for the distribution of sediments on the continental shelf, the processes that affected and are affecting these sediments, and the paleoceanographic record of southeastern Canada. Regional sediment mapping programs undertaken in the late 1960's and 1970's relied heavily on textural information obtained from numerous seabed grab samples (Fig. 1) and geomorphological and acoustical information interpreted from echosounder profiles. These studies resulted in the production of a suite of surficial sediment maps and the identification of a widespread low sea-level position of -110 to -120m as defined by the boundary between two surficial formations, the Sable Island Sand and Gravel and the Sambro Sand (King, 1967, 1970, Fig. 2). Subsequent dating of the low sea-level stand (King and Fader, 1986), at 15 ka BP, provided the chronostratigraphic control. For the eastern and western areas of the Scotian Shelf, the textural and acoustic evidence was not published. The emphasis of the mapping program was to characterize the sediments in terms of surficial formations, and the detailed textural, lithologic and acoustic information was consolidated as facies within the formational concepts. As a result, the interpreted low sea-level position has been questioned by such studies as those of Quinlan and Beaumont (1981, 1982) who have modelled crustal response to glacial loading during the Late Pleistocene. Results of those studies have predicted a maximum lowering of sealevel of 70m for the outer continental shelf. Also, Scott and Medioli (1982) have presented evidence of a maximum lowering of sealevel of only 30m below present along the Nova Scotian shoreline.

Additional quantitative and qualitative information on the low sea-level stand has been provided by the development and application of new marine technology in the late 1970's and 1980's, such as rectified sidescan sonars, high resolution seismic reflection systems and their associated seabed reflectivity coefficients, submersibles, remotely controlled camera vehicles, towed sleds and industry drilled boreholes. The surficial mapping program has expanded to the Grand Banks of Newfoundland from the adjacent Scotian Shelf (Fig. 3) and to date over 5000 samples have been collected and analysed (Sonnichsen et al., 1987). The methodology has changed since the first areas were studied and emphasis is now placed on the acquisition of fewer samples but more high quality high resolution seismic reflection and sidescan sonar data. The development of remote sediment classification methods using signal processing techniques has led to the collection of even fewer samples to calibrate the acoustic information (Parrott et al., 1980). Using digital seismic reflection data collected with a broad band boomer source, reflected energy is measured from the top metre of seabed. Acoustic reflectivity is calculated as a percentage of the energy of the original source pulse. The surficial formations display a characteristic range of reflectivity values which provides a lithologic component to the stratigraphy normally displayed on the profiles. Table 1 is a summary of acoustic reflectivity measurements for the surficial formations from the Scotian Shelf.

GEOMORPHOLOGY

Terraces were first recognized on the Scotian Shelf in the early 1960's by King (1967), Stanley et al., (1968), and James and Stanley (1968), who interpreted them to represent features formed by "surf base" erosion in former nearshore environments. They are described as relatively flat or gently inclined surfaces which are bounded on one side by a steeper ascending slope and by a descending slope on the opposite side. The uniformity of

74

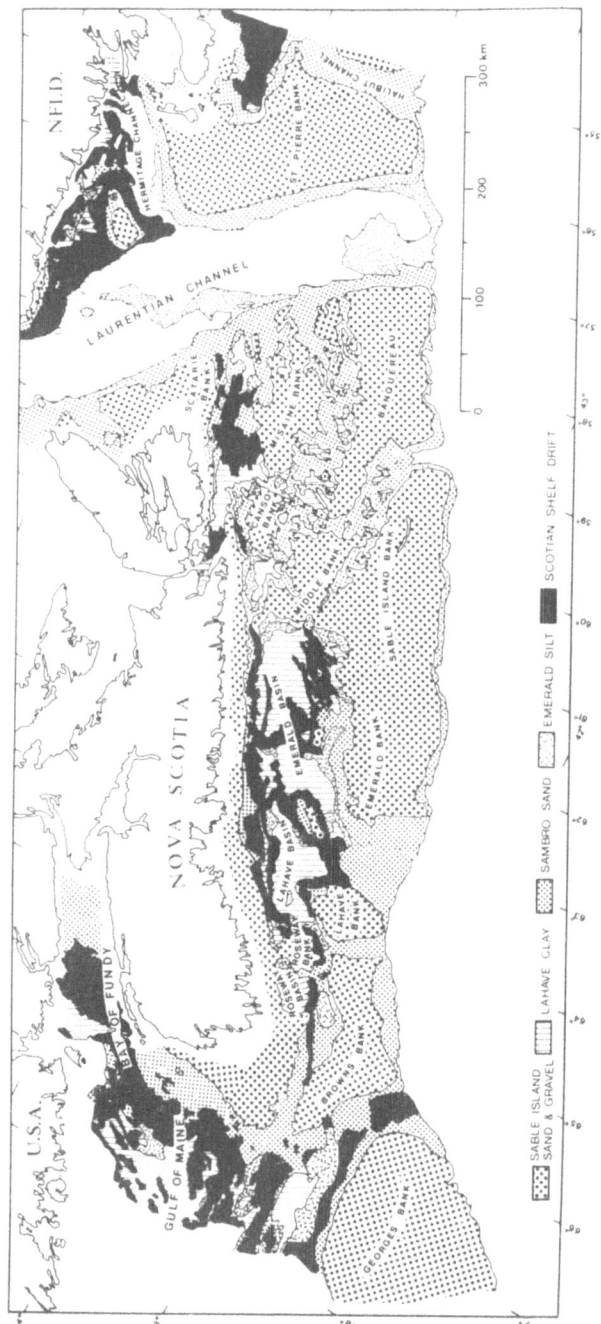

2. Distribution of surficial formations across the Scotian Shelf from the Gulf of Maine to the Grand Banks of Newfoundland, from King and Fader (1986). The boundary between the Sambro Sand and the Sable Island Sand and Gravel formations occurs at the late Wisconsinan low sea-level stand of 110–120m.

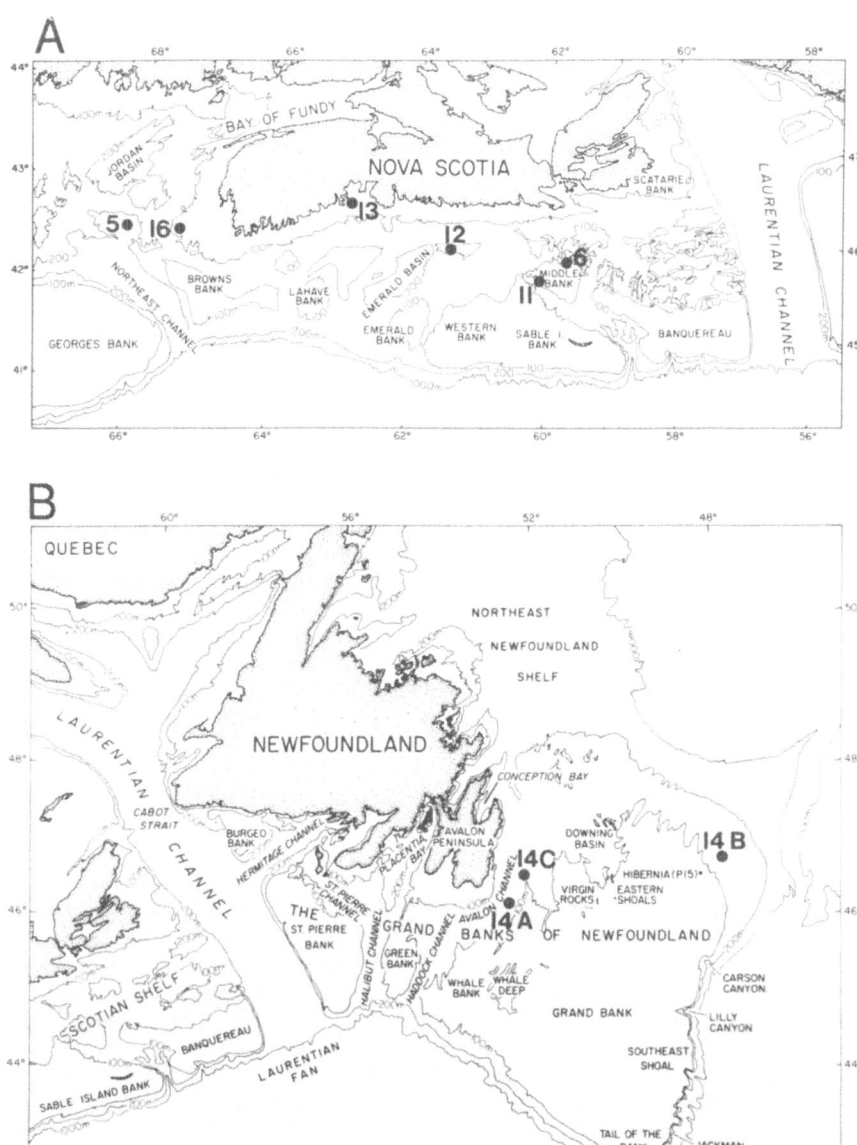

3. Index for the Scotian Shelf and the Grand Banks of Newfoundland. The surficial geology of the Scotian Shelf has been mapped and the Grand Banks is presently under study. The location of seismic profiles and sidescan sonograms illustrated in the paper are shown.

terrace depths strongly suggested an origin related to former sea levels. Figure 4, from Stanley et al., 1968, is a histogram showing the depth and frequency of the most prominent terraces on the Scotian Shelf interpreted from a study of echograms. On the basis of petrographic examination of sediments collected from the terraces and comparison with the depths of other terraces found on other sectors of the Atlantic continental margin, they interpreted a 145m terrace as Illinoian or earlier in age and a 120m terrace as representing the minimum stand of sea level during the last Wisconsinan glacial advance. They also recognized the shortcomings in interpreting terraces based only on echogram data and suggested that terraces may represent subaerial erosional features, bedrock benches, fault planes, slump scars or features formed through glacial erosion. We now can evaluate these interpretations through the application of high resolution seismic reflection and sidescan sonar technology. The areas of greatest inaccuracy in interpreting terraces as sea-level indicators occur where: 1) sand ridges and other large scale bedforms are present, 2) where relict and modern populations of iceberg furrows cover the seabed, 3) where slumped sediments occur, particularly on the seaward flanks of the outer shelf banks, and 4) where changing bedrock lithology or structure gives rise to terrace-like features. These features and processes often obscure the original morphology, sediment distributions and stratigraphy, making interpretations on low sea level difficult, or they mimic terraces formed by low sea-level stands. The earlier recognized terraces were thought to be mainly cut in bedrock but the new data show that most are developed across surficial sediments. King and Fader (1986), in a regional study of the glacial sediments across the shelf, have identified a blanket deposit of surficial sediments, approximately 50m in thickness, covering the bedrock surface. The only extensive areas of exposed bedrock occur on the inner shelf. Terrace-like features can also result from erosional processes where currents have removed sediments, for example in Figure 5, where a gravel-rich layer known regionally in the Gulf of Maine as the Truxton event has been exposed within the glaciomarine section and forms a prominent shoulder, or where depositional processes have built sand ridges across flat gravel lag deposits, which appear in cross-section as notch-like features.

During the mapping of the surficial sediments across the Scotian Shelf, many terrace-like features were recognized on the echogram data, but until the development of high resolution seismic reflection systems such as the Huntec DTS system (Hutchins et al., 1976), the stratigraphy of the terraces remained unresolved. Figure 6 is a Huntec high resolution seismic reflection profile across a terrace at a water depth of 110-120m on the northern edge of Middle Bank on the Scotian Shelf. It appears as a bench, cut in both Tertiary bedrock and glacial sediments, and is overlain by flat-lying reflections interpreted as sand. Echosounder systems would not penetrate the sand and gravel sediment on this terrace and define the stratigraphy.

VISUAL OBSERVATIONS

Visual observations of the terrace were made by Amos (1987), using the submersible Pisces IV on the flanks of Sable Island Bank and Banquereau. In his Zone 7 of Dive 1609, in a feeder canyon on the edge of Sable Island Bank, well-rounded boulders occurred in a band concentrated at the 110m isobath. This is interpreted as a region of intense sediment reworking based on an alignment of the boulders and the occurrence of compacted horizons of well-sorted gravel on a clearly definable terrace cut in glacial till. Below the terrace, the gravel sized fraction, which consisted of large boulders and pebble sized material, was angular in shape and the associated finer grained sediment was poorly sorted.

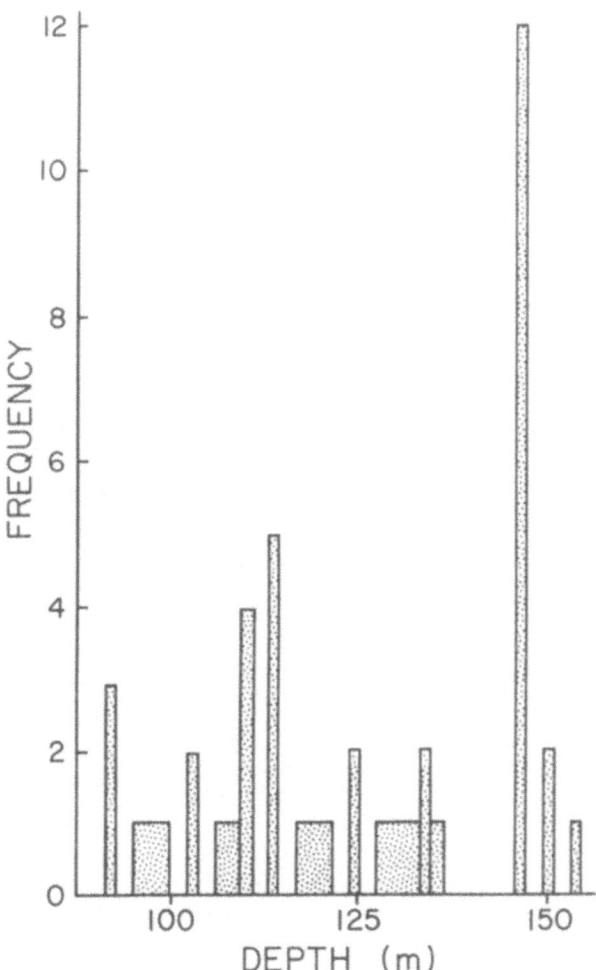

4. A histogram showing the depth and frequency of terraces on the Scotian Shelf from Stanley et al. (1968). The deepest terrace at 146m occurs on the seaward margins of the outer shelf banks and was thought to be older than Wisconsinan in age. The histogram was compiled from echosounder data mainly from the Western and Sable Island Bank areas of the shelf. The mode at 110-120m is the low sea-level stand as interpreted in this paper.

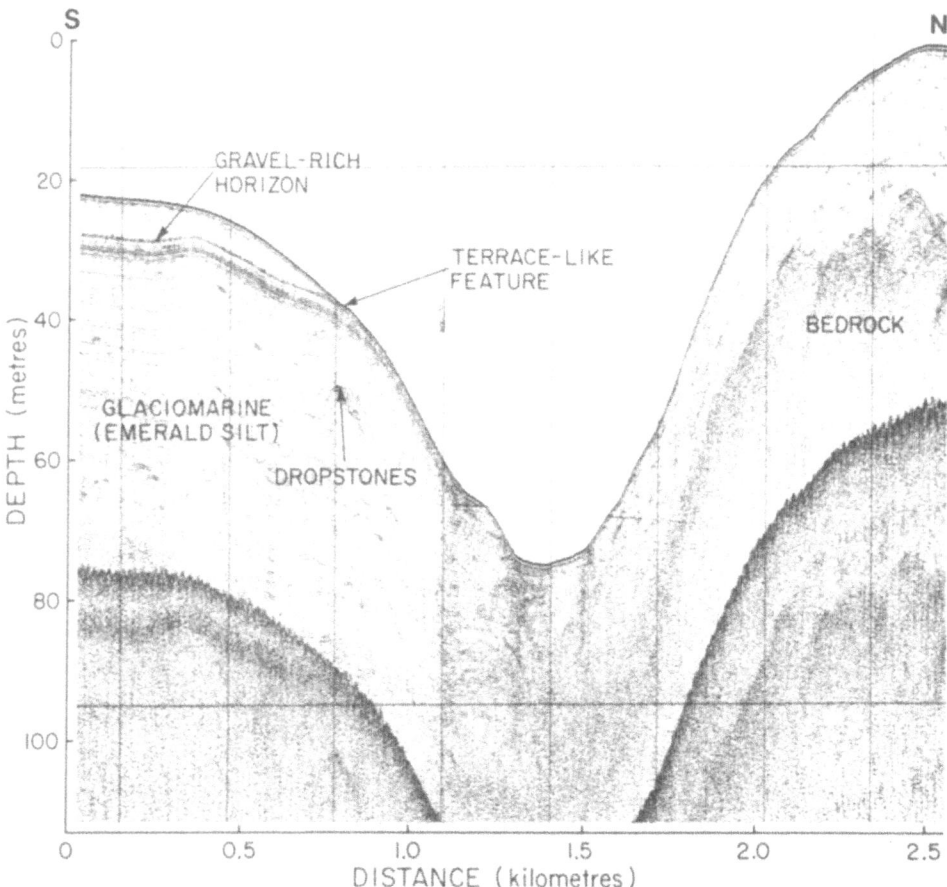

5. A Huntec high resolution seismic reflection profile from the Gulf of Maine showing a terrace-like feature at the seabed in 220m water depth. Using the seismic reflection data and sediment cores from the area the feature is defined as a gravel rich horizon within muddy glaciomarine sediments. Postglacial erosion of the overlying muddy sediments by strong currents has exposed the gravel bed giving the seabed a terrace-like appearance.

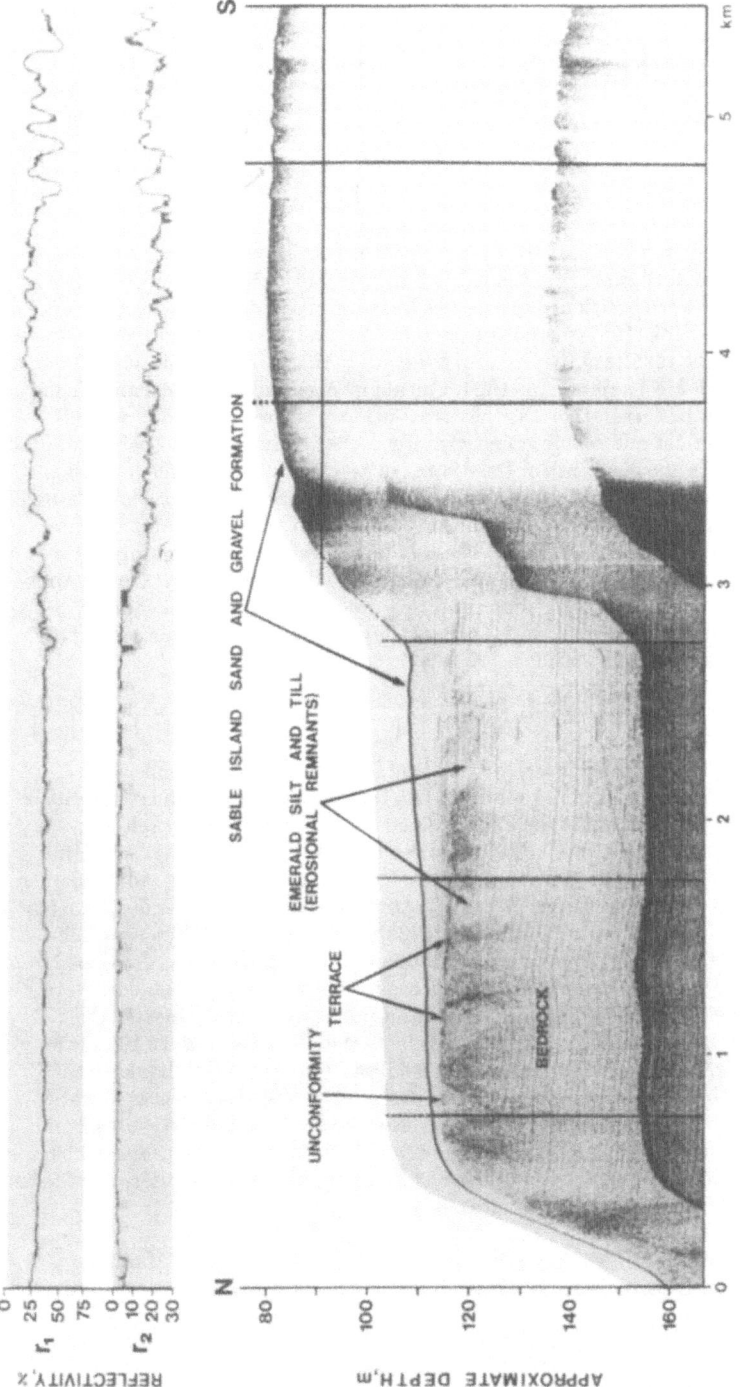

6. A Huntec DTS profile across the 110-120m terrace on the northern edge of Middle Bank, Scotian Shelf. The terrace is interpreted to be cut into Tertiary bedrock and glacial sediments and is overlain by a deposit of sand and gravel. Echosounder data does not penetrate the sediments and define the stratigraphy. Profile courtesy of D. B. Scott, Dalhousie University.

A five metre high scarp on the southwest flank of Banquereau at a depth of 120m was also investigated with the submersible (Amos, 1987). Previous interpretations suggested that the scarp was a wave-cut terrace (King, 1970). Observations on the distribution of well-sorted gravel, a lack of features indicative of sediment motion such as ripples, and the scarp itself, led Amos to suggest that the feature represents the late Wisconsinan low sea-level stand.

TEXTURAL EVIDENCE

The Halifax-Sable Island map area was the first area of the Scotian Shelf to be sampled systematically to groundtruth echogram acoustic signatures. In addition, many closely spaced sample transects were positioned to extend across the suspected depth of occurrence of the low sea-level stand. King (1967) presented the textural analysis on 141 samples from the area and in comparing the results of the textural analyses to the acoustic signatures, found a direct correlation. Table 2 is a summary of the textural and acoustic characteristics. The most striking feature of the correlation was the distribution of the well-sorted sands and rounded clasts in the boulder and cobble range with the acoustically hard, non-penetratable seabed and its occurrence above a depth of 110-120m. In depths immediately below this level, the sediments contained up to 20% silt and clay-sized particles. Often the low sea-level position could be identified by whether or not the sea water within the grab sampler was muddy or was free of suspended particulates. The clay facies were confined mainly to the deeper basins surrounded by the silty sediments. Adjacent to the sandy and gravelly banks, the sediments were dominantly muddy sands. Figure 7 shows the grain size frequency curves for the Sable Island Sand and Gravel, LaHave Clay, Sambro Sand, Emerald Silt and the Scotian Shelf Drift surficial formations.

Particle Shape

Shape analysis, particularly applied to the gravel-sized fraction, is another important parameter in determining the position of a low sea-level stand. Roundness and sphericity of gravel clasts (pebble to boulder range) were determined using bottom photographs, submersible observations, and samples collected with a large volume clam-shell sampling device. From a study of boulders on the Grand Banks, Fader and Miller (1986) and Fader (in press), reported a large population of rounded to subrounded boulders in water depths of less than 100m (Figure 8). In deeper water extending to 220m, most of the boulders were sub-angular to angular in shape. They concluded that the rounding (Figure 9) was developed in the beach or surf zone of the late Pleistocene transgressing sea and that the boulders were transported to the shelf by glaciers. Beneath large sand ridges which cover most of the surface of Grand Bank, a pavement of well-rounded clasts is ubiquitous (Figure 10). As on the Scotian Shelf, the gravel is often continuous, extending from the shallow banks to the depressions of the shelf, but the clasts greatly differ in shape. A transition occurs in water depths of 110-120m. Above these depths the pebbles and cobbles are often disk-shaped and well-sorted suggesting formation in a high energy beach environment. In the adjacent basins of the shelf the pebble-cobble sized clasts are angular to subangular and are interpreted to be sorted through iceberg furrowing (Fader, in press).

In contrast to the Scotian Shelf, the low sea-level stand in the Eastern Gulf of Maine is warped up to the north and decreases in depth from 110-120m on Brown's Bank to less than

TABLE I: RESULTS OF ACOUSTIC REFLECTIVITY MEASUREMENTS FOR THE SURFICIAL FORMATIONS ON THE SCOTIAN SHELF.

Geological Formation	Acoustic Reflectivity						Interpretation of Energy Content			
	Range of Trend		Typical Values		Variability about Trend		Surface Reflectivity		Internal Reflectivity	
	r_1	r_2	r_1	r_2	r_1	r_2	Coherent	In-coherent	Coherent	In-coherent
LaHave Clay	5-10	3-5	8	3	Low	Low	Low	Low	Low	Low
Emerald Silt	10-25	5-12	17	7	Low-Medium	Low-Medium	Medium	Low	High	Medium
Scotian Shelf Drift (glacial till)	20-50	7-12	30	10	Medium-High	Medium-High	Medium	Low-High	Low	Medium
Sambro Sand	30-50	5-10	40	7	Low	Low	High	Low	Low	Low
Sable Island Sand and Gravel Gravel Facies	35-60	10-25	50	15	High	High	High	High	Low	High
Sand Facies	35-50	5-10	40	6	Low-Medium	Low-Medium	High	Low	Low	Low-Medium
Megarippled Sand	25-35	15-25	30	20	Medium	Medium	Medium	Very High	Low	Low
Bedrock (meta-morphic-igneous)	10-40	10-30			High	High	Low-Medium	Very High	Low	Low

Notes: The reflectivities, expressed as a percentage, show ranges and typical values from the analysis of 250 km of seismic reflection data on the Scotian Shelf. The distribution of energy content is based upon an interpretation of reflectivity profiles, and the character of the seismic section.

TABLE II: CONDENSED DESCRIPTION OF TEXTURAL PROPERTIES AND ECHOGRAM INTERPRETATIONS FOR THE SEDIMENTARY FACIES

Sedimentary Facies (Formation)	Average Textural Grades (%)				Average Median Diam. mm	Echogram Interpretation	Remarks
	Gravel	Sand	Silt	Clay			
Marine and Transitional:							
Sand and Gravel (Sable Island)	47	53				Type IV hard, smooth.	Hard bottom above submarine terrace and coarse textural grade define the unit.
Sand (Sambro)	15	68	13	4	0.23	Type IV hard, smooth bottom.	Hard bottom below submarine terrace and near textural grade with respect to the sand and gravel facies define this unit.
Silt (Emerald)		30	47	23	0.043	Type II semi-compacted bottom with smooth surface.	The smooth surface and the lack of gravel are the significant criteria in distinguishing it from glacial till.
Clay (LaHave)		3	39	58	0.003	Type I soft, smooth bottom	Acoustical transparency and high clay content are the chief diagnostic features.
Continental:							
Glacial Till (Scotian Shelf Drift)	3	64	17	16	0.117	Type III, undulating semi-compacted to hard bottom	Undulating bottom, wide range of textural grades are the most significant characteristics.

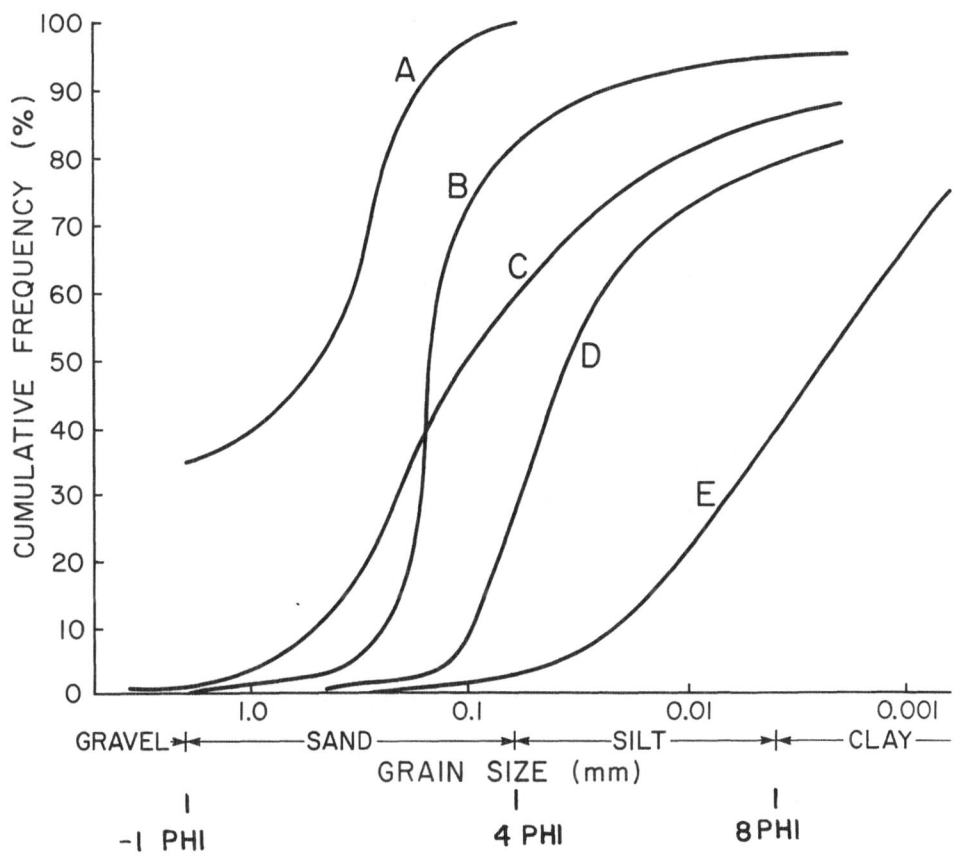

7. Grain-size frequency distribution curves for samples of a) Sable Island Sand and Gravel, b) Sambro Sand, c) Scotian Shelf Drift, d) Emerald Silt, and e) LaHave Clay (after King, 1966). Approximately 150 seabed samples were used in the compilation.

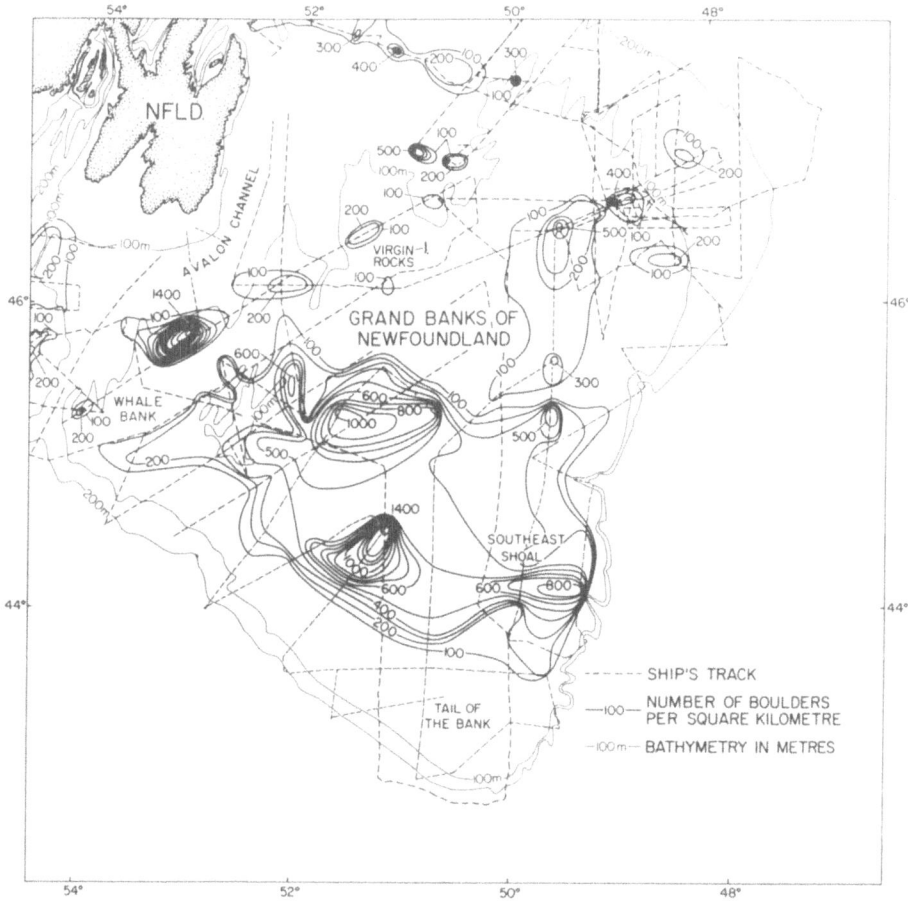

8. Distribution of boulders larger than 0.5m across the eastern Grand Banks of Newfoundland interpreted largely from 100khz sidescan sonograms.

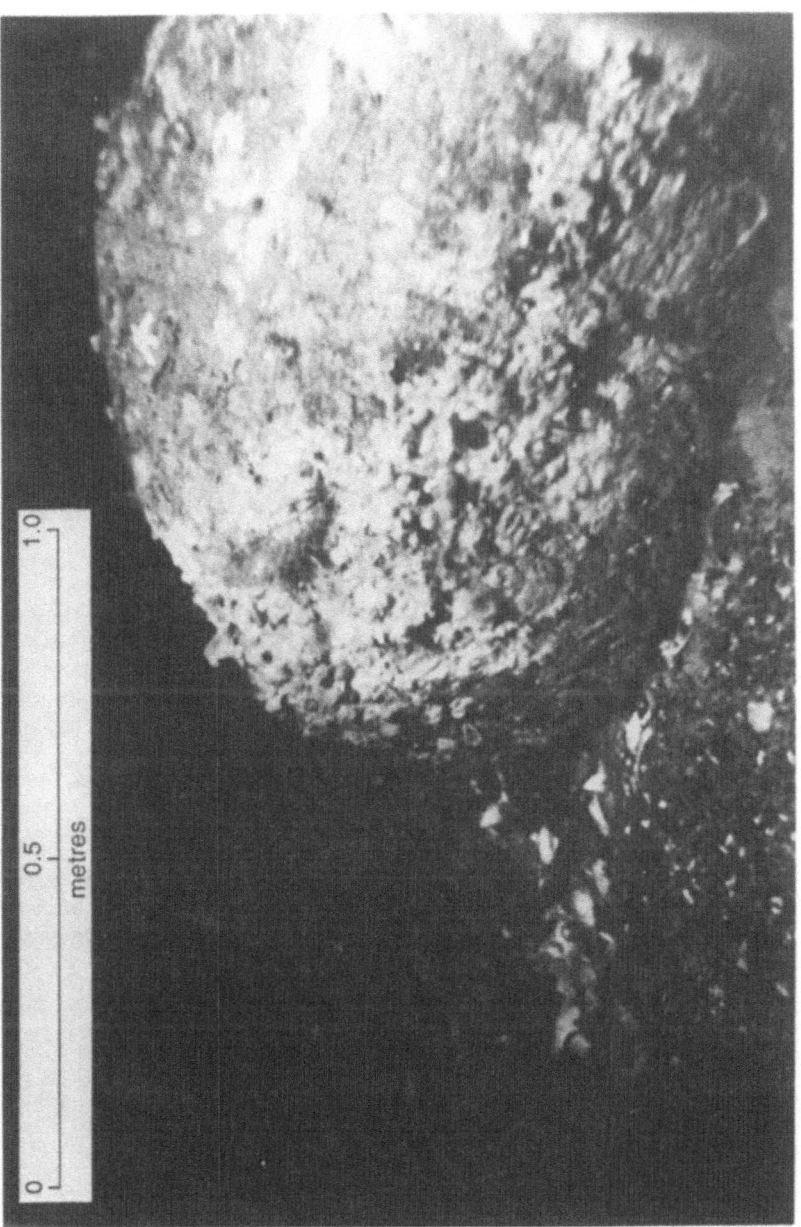

9. Seabed photograph of a 1.5m diameter rounded boulder from the southern area of Grand Bank. This boulder lies on a pavement of well-rounded cobbles and pebbles in a trough area of large sand ridges.

10. A typical example of rounded and disk-shaped gravel at the seabed in areas above the low sea-level stand of 110-120m. The sorting and shaping is attributed to littoral processes during the Late Wisconsinan transgression. Photograph from St. Pierre Bank, south of Newfoundland.

37m at the entrance to the Bay of Fundy (Fader et al., 1977). Numerous, soft, angular fragments of Tertiary mudstone are found in depths as shallow as 92m in the area and could not have survived reworking by a transgression. Their distribution indicates that the low sea-level stand occurs at a lesser depth. Late glacial rebound of the crust in the Fundy-northern Gulf of Maine region is responsible for the difference in elevation of the terrace and is in agreement with recent studies on the late Wisconsinan glacial history of coastal and nearshore Maine (Belknap et al.,1986). This example illustrates the usefulness of particle shape and lithology as parameters in determining the limits of the low sea-level stand.

STRATIGRAPHIC EVIDENCE

Acoustic investigations were undertaken to map the distribution of sediments, especially terrace gravels, and to determine stratigraphic relationships among surficial sedimentary units, terrace morphology and shelf unconformities. The initial acoustic investigations of the Scotian Shelf were undertaken with a Kelvin Hughes MS 26B echosounder system and the methods for interpretation are summarized in King (1980). The distribution of all acoustically hard and flat seabeds appeared confined to water depths above 110-120m. Below 140m water depth, penetration of the acoustic energy increased indicating finer grained seafloor sediment. The limiting aspect of echograms in penetrating hard seabeds was overcome with the development of the Huntec DTS system (Hutchins, 1974 ; McKeown, 1975; Hutchins et al., 1976). This system provided structural details within the surficial and upper bedrock units and penetrated up to 50m in harder sediments such as tills and Tertiary sandstones. In depths above the low sea-level stand, widespread unconformities occur that are developed across tills, glaciomarine sediments and bedrock (Fig. 11) . These sediments often exist as erosional remnants with thin deposits of transgressive sand and gravel overlying the unconformity. In contrast to the eroded and discontinuous glacial sediments on the banks of the Scotian Shelf, within the basins the glaciomarine sediments are thick, continuous, conformable and generally without structural disturbances (Fig. 12). Unconformities within the major basins are rare, but where they occur they can be attributed to current erosion that developed during a low sea-level stand or perhaps local, late glacial erosion on the basin flanks. Such areas appear mainly confined to the zone of dissected topography to the north of Banquereau and on the inner shelf associated with morainic ridges. The distribution of these unconformities and glacial remnants supports the interpretation that indeed a transgression has eroded the glacial sediments in water depths less than 110m.

Distribution of Till

The distribution of till on the inner shelf is another important factor in determining the low sea-level stand. Examination of air gun and Huntec DTS profiles and sidescan sonograms on the inner shelf extending offshore from the headlands to 110-120m water depth, shows the character of the seabed to be largely that of exposed bedrock over much of the area, as typified by the sidescan sonar image in Fig. 13. However, thicker deposits of till remain in buried bedrock channel systems which are interpreted as drowned rivers. This absence of till on the inner shelf is interpreted to result from erosion of previously deposited till and glaciomarine sediment by a transgressing sea. The surface of the bedrock and seabed is more steeply dipping in this area of the shelf and undercutting, slumping and subaerial freeze-thaw mechanisms are proposed to account for the erosion. Along the eastern shore of Nova Scotia, this process is presently active and till cliffs over 30m in height are rapidly eroding.

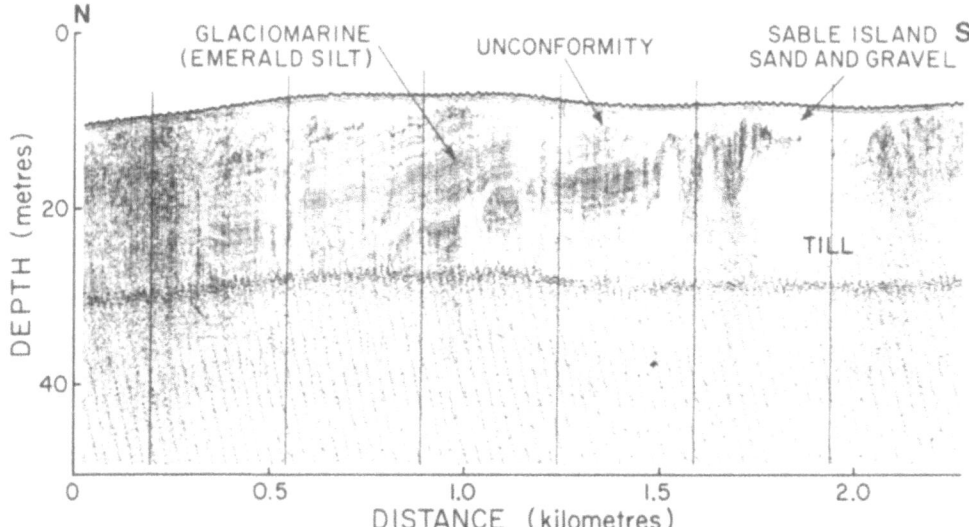

11. A Huntec DTS profile from Middle Bank, Scotian Shelf, showing erosional remnants of Emerald Silt with a well-developed unconformity across their surface. Overlying the Emerald Silt is a thin deposit of Sable Island Sand and Gravel, up to 3m in thickness. This stratigraphic relationship is common across the shelf in water depths above the low sea-level stand, and is in sharp contrast to the deeper basinal areas where the sediments are thicker, conformable and not eroded (Fig. 12).

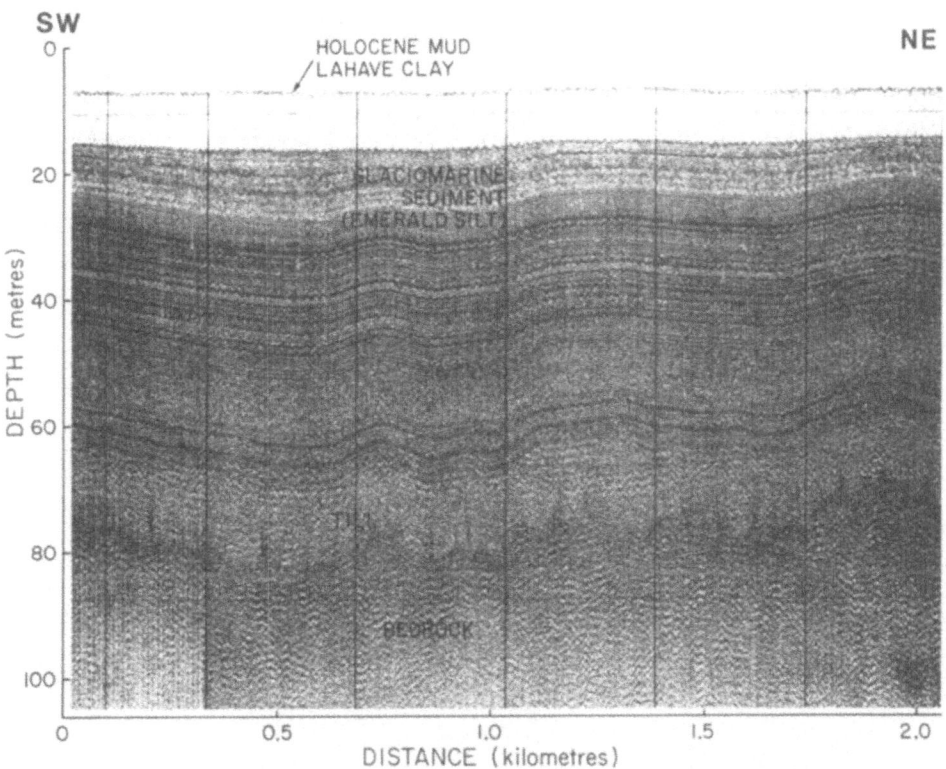

SW   HOLOCENE MUD   NE

12. A Huntec DTS profile from Emerald Basin, Scotian Shelf, illustrating a typical basin stratigraphy of conformable, thick, glaciomarine sediments and an absence of internal erosional unconformities.

90

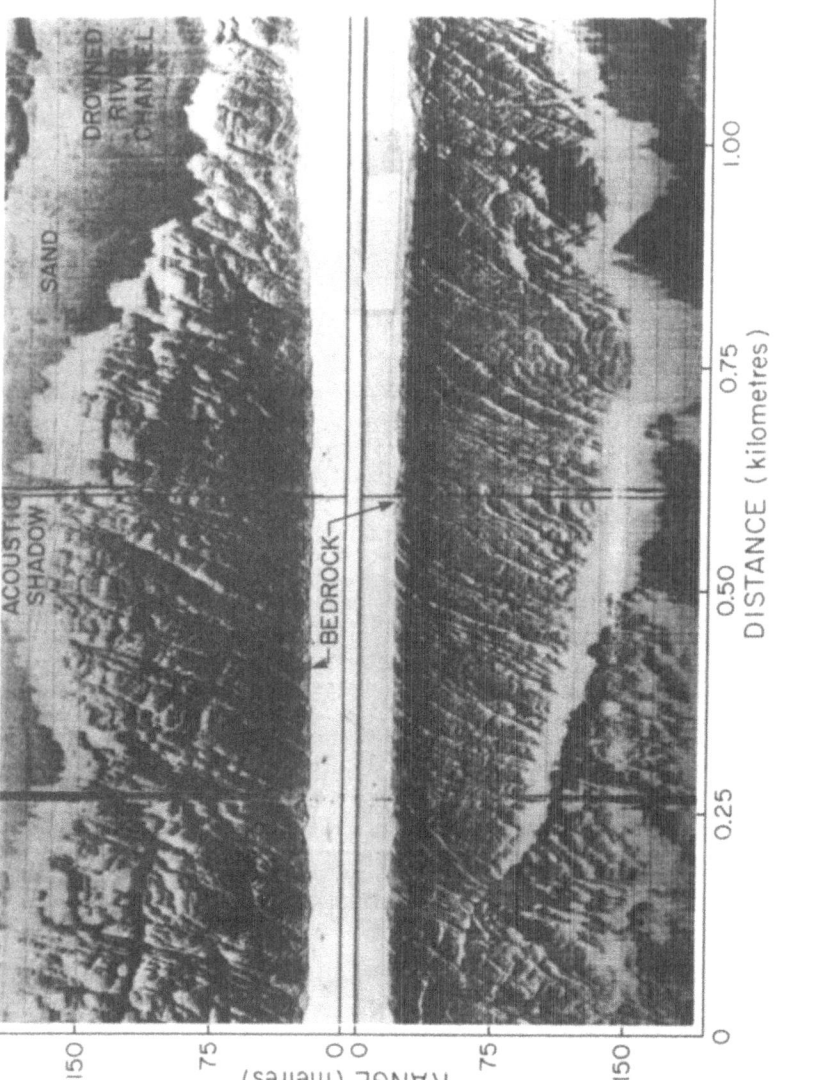

13. Sidescan sonogram from the inner part of the Scotian Shelf showing exposed bedrock at the seabed overlain by thin, patchy, sand and gravel. The absence of a glacial section is typical of the inner shelf above 110-120m and is attributed to erosion of preexisting till and glaciomarine sediment during the transgression. Sonogram courtesy of Gedoro Co. Ltd.

In the protected coastal embayments and harbours of Nova Scotia, erosion also occurs but to a lesser degree. The deposits of till are only partially removed as they quickly become armoured with coarse gravel, preventing complete erosion (Letson,1980).

ICE-SEABED EVIDENCE

The distribution of iceberg furrows and pits on the seabed which result from the grounding of drifting icebergs is another characteristic that provides information on the position of the low sea-level stand. The Grand Banks of Newfoundland is approximately the southern limit of occurrence of modern, seafloor scouring icebergs. On the Scotian Shelf iceberg furrows are only found on till (King, 1980; King and Fader, 1986), and are interpreted as a relict population. On the Grand Banks, Fader and King (1981) recognized two iceberg furrow populations (Fig. 14), and interpreted their distribution in relation to the low sea-level stand of 100m. Population 1 consists of a deeper zone of unmodified iceberg scours with a shallower eroded and infilled zone. The deeper zone occurs in water depths generally greater than 150m and is characterized by a seabed completely scoured by icebergs which manifests itself on the sidescan sonograms as criss-crossing patterns of linear ridges and interspersed trough areas (Fig.14 a). Isolated, circular-lenticular depressions (iceberg pits) resulting from vertical iceberg-seabed loading are also common. The shallower zone of population 1 occurs in depths ranging between 130-150m and appears on the sonograms as a network of discontinuous lineations. It is interpreted as an eroded and partially infilled iceberg furrowed seabed where all that remains of the iceberg furrows are discontinuous bouldery rims (Fig. 14b). Population 2 is limited in occurrence to the northern areas of the Grand Banks of Newfoundland in water depths less than 200m. The seabed is characterized by a sparse population of iceberg furrows that exhibit well defined berms which may be described as "fresh" in appearance (Fig. 14c). The deep water dense distribution of iceberg furrows of population 1 grades gradually upslope in shallower water into the eroded infilled zone on the bank flanks, and ends abruptly at 100-110m water depth. Only population 2 furrows occur above 100m water depth.

This abrupt iceberg furrow boundary is interpreted to closely approximate the low sea-level stand on the northern Grand Banks. The eroded and partially infilled iceberg furrows are interpreted to have been modified by currents and waves in the sublittoral zone of the low sea-level stand with the degree of modification decreasing with depth until the unmodified population 1 is reached in deeper water. The deeper scour population could have once extended up onto the bank but these would have been effaced by erosion during the subsequent low sea-level cycle. The lower limit of erosion in this cycle at the lowstand trimmed the upslope limit of the older ice scour population to just below 110m and degraded the adjacent remaining scour marks. As erosion occurs to the base of the shoreface, or 2-3 times the maximum height of breaking waves (Bruun, 1983), the abrupt upper boundary of the relict ice scour population is interpreted as the surf base or level of the lower shoreface below the lowest stand of the late Wisconsinan sea level. From a review of eroding coasts world-wide, King (1972) ascribes a general value of about 10m below sea level for the depth of active erosion, but on the wave exposed eroding eastern shore of Nova Scotia, Boyd and Penland (1984) and Forbes (1986), commonly find the base of the shoreface at the 15-20m isobath. Thus the interpreted sea level on northern Grand Bank is about 90-100m on the basis of evidence of an eroded relict iceberg scour population. This distribution also helps constrain the ages of the iceberg furrows on the banks of the Grand Banks. If the eroded population on the flanks of the banks was modified in the adjacent shallow water of the low

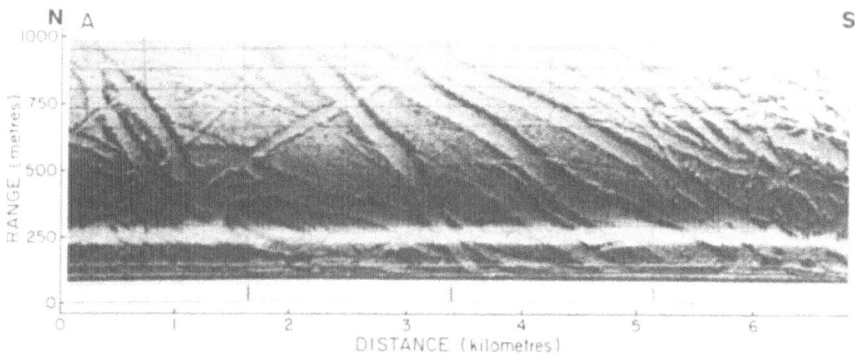

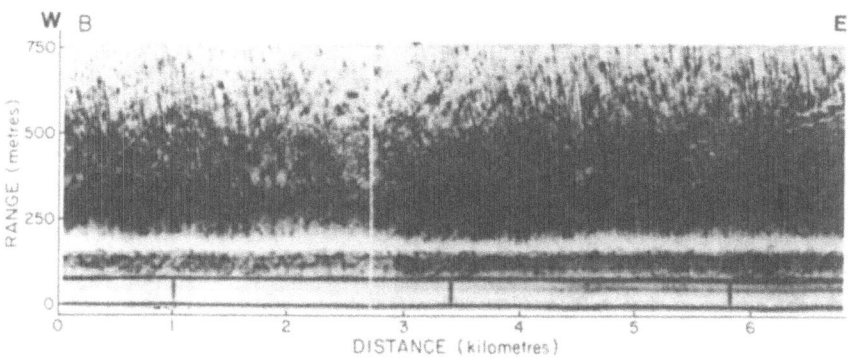

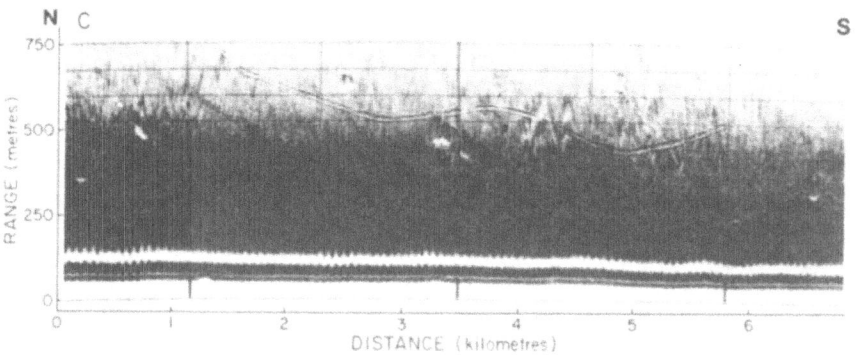

14. Sidescan sonograms from the Grand Banks of Newfoundland showing two populations of iceberg scours: a) an intensely furrowed seabed in 200m water depth of population 1, b)population 1 in an area slightly below the low sea-level stand at 140m showing eroded and infilled furrows and c) population 2 from Grand Bank in 90m water depth showing modern, fresh looking iceberg furrows. Population in (b)is interpreted to have been eroded in the sublittoral zone of the Late Wisconsinan low sea-level stand, having formed earlier when sea level was higher.

sea -level stand, then population 2 on the banks must postdate the subsequent transgression. A sufficient depth of water is also necessary to carry large enough icebergs to scour the coarse gravelly materials on the banks. If we accept the age of the low sea-level stand as approximately 15 ka BP, then iceberg furrow population 2 must be younger (Fader and King, 1981).

Population 2 of iceberg furrows is also found superimposed on the degraded relict population below 110m. This population is attributed to the Late Holocene (Lewis et al., 1987), on the basis of ice-rafting evidence and computer modelling of grounding frequencies of present and historical iceberg flux.

GEOTECHNICAL AND GEOCHEMICAL EVIDENCE

From a study of geotechnical boreholes through Tertiary sediments and vibrocores of Pleistocene-Holocene sand ridges on northern Grand Bank, Segall et al. (1987) have provided geotechnical and clay mineral evidence in support of a subaerial exposure and subsequent transgression of the northeast Grand Banks of Newfoundland. Based on mineralogical studies of clay-sized material, a weathered desiccated crust at the Tertiary unconformity has been identified, which overlies sediments deposited in an alkaline marine environment under more normal marine conditions. The Pleistocene-Holocene sands and gravels overlying the Tertiary section are interpreted to have been derived from a northern latitude provenance.

Additionally, shells are absent from the weathered zone while present in the overlying Holocene sediments and in the deeper underlying Tertiary section. Shell casts have been identified from the weathered zone. Drill logs maintained during collection of the boreholes indicated great difficulty in penetration at the Tertiary unconformity. Analysis of a borehole collected in slightly deeper water just above 110m shows a thinning of the weathered zone and the presence of shells throughout the entire section. These data provide control on the minimum depth of occurrence of the subaerially exposed Tertiary sediments, hence a low sea-level stand, and suggest that it occurred at least in depths of 100m. However, the geochemical data do not provide age control on the timing of the low sea-level position. In addition to the dating of shells in the overlying sands and gravels, which range in age from approximately 3 -10 ka BP, dating of glacial sediments which exist as erosional remnants on the Grand Banks in adjacent buried channel systems will provide the chronostratigraphic control. The depth of the low sea-level stand is, however, consistent regionally with the dated low level from adjacent areas. The similar distribution and structural characteristics of transgressed sediments support the interpretation that for the northern Grand Banks of Newfoundland, the low stand occurred during the late Wisconsinan.

TERRACE AGE

King (1965) dated shells collected at depths of between 99 and 117m within the Sable Island Sand and Gravel formation above the interpreted low sea-level stand to provide age control. These dates ranged in age from 8708 to 9990 a BP and suggested that the low sea-level stand was older. It is difficult to determine the age of the low sea-level stand as conditions at the time of its formation and the subsequent transgression have resulted in an overconsolidation of the sediments and deposition of coarse sand and gravel, which is difficult to penetrate with conventional piston cores. Most of the cores from the Scotian Shelf have therefore been collected in the basinal areas and have sampled the glacial and postglacial muddy facies. However, on the northeastern flank of Emerald Basin near the low sea-level stand two piston cores penetrated glaciomarine sediments which were eroded in the beach zone of the trangressing sea and are overlain by postglacial clay (King and Fader, 1986). The youngest glaciomarine date of 15 .1 ka BP in core HN-79011-1 and the oldest date of 14.465 ka BP in core DW-82003-4 s from the overlying postglacial sediment, bracket the age of the low sea-level stand. Additional cores from Emerald Basin (King and Fader, 1986), collected within the glaciomarine section, were dated to 41.8 ka BP and indicate that glacial conditions persisted across the shelf until 15 .1 ka BP.

From a study of industry borehole samples and regional and site-survey seismic profiles on Banquereau, Amos and Knoll (1987) provide additional age control on the low sea-level stand and subsequent transgression. They interpret the age of the low sea-level stand at approximately 14 ka BP based on the distribution of subaerially formed channel gravels above 120m water depth and an unconformity developed across widespread deposits of glaciomarine sediments dating from 16 ka BP to 28 ka BP. These are buried beneath deposits dating 1.5-8 ka BP. Figure 15, from Amos and Knoll (1987), shows a plot of ages against depth below sea level on which their study was based. These dates, together with the those of King and Fader (1986), provide regional age control on the low sea-level stand and indicate formation at approximately 15 ka BP.

VARIATIONS IN THE LOW SEA-LEVEL POSITION

How sensitive are these seabed and sedimentological characteristics in determining the low sea-level position and do these interpretations agree with other evidence on sea levels from the adjacent land areas? To address this question it is useful to examine areas where in the offshore the sea-level position appears to be warped or occurs at different elevations. Such areas occur in the Gulf of Maine at the entrance to the Bay of Fundy, in Chedabucto Bay south of Cape Breton Island, adjacent to the south coast of Newfoundland, and on the southern Grand Banks of Newfoundland on the Tail of The Bank.

Outer Bay of Fundy

As discussed earlier, in the approaches to the Bay of Fundy the presence of angular fragments of Tertiary mudstone in progressively shallower water to the north first indicated that the low sea-level position decreased in water depth (Fader et al., 1977). In addition, the percentage of silt and clay-sized sediment within the seabed samples increased in a similar fashion northward. The gravel sized fraction on the floor of the Bay of Fundy was angular to subangular in shape and indicated that the material had not been transgressed. Thick accumulations of till in water depths as shallow as 37m also suggested a higher position for the low sea-level stand. Farther up the Bay of Fundy at the mouth of Chignecto Bay, Scott and Greenburg (1983) show a low sea-level stand of 35m depth at 7 ka BP.

For the Fundy region the sedimentologic, lithologic and seismostratigraphic evidence supports the interpretation of a northward shallowing of the low sea-level stand from 110-120m on Brown's Bank to less than 37m within the Bay of Fundy. This interpretation is in agreement with the terrestrial evidence from the area which consists of 1) a marine limit in southwest Nova Scotia at a maximum elevation of 45m, dated at approximately 14 ka BP (Grant, 1971), and 2) nearshore coastal and raised marine deposits of southern Maine which

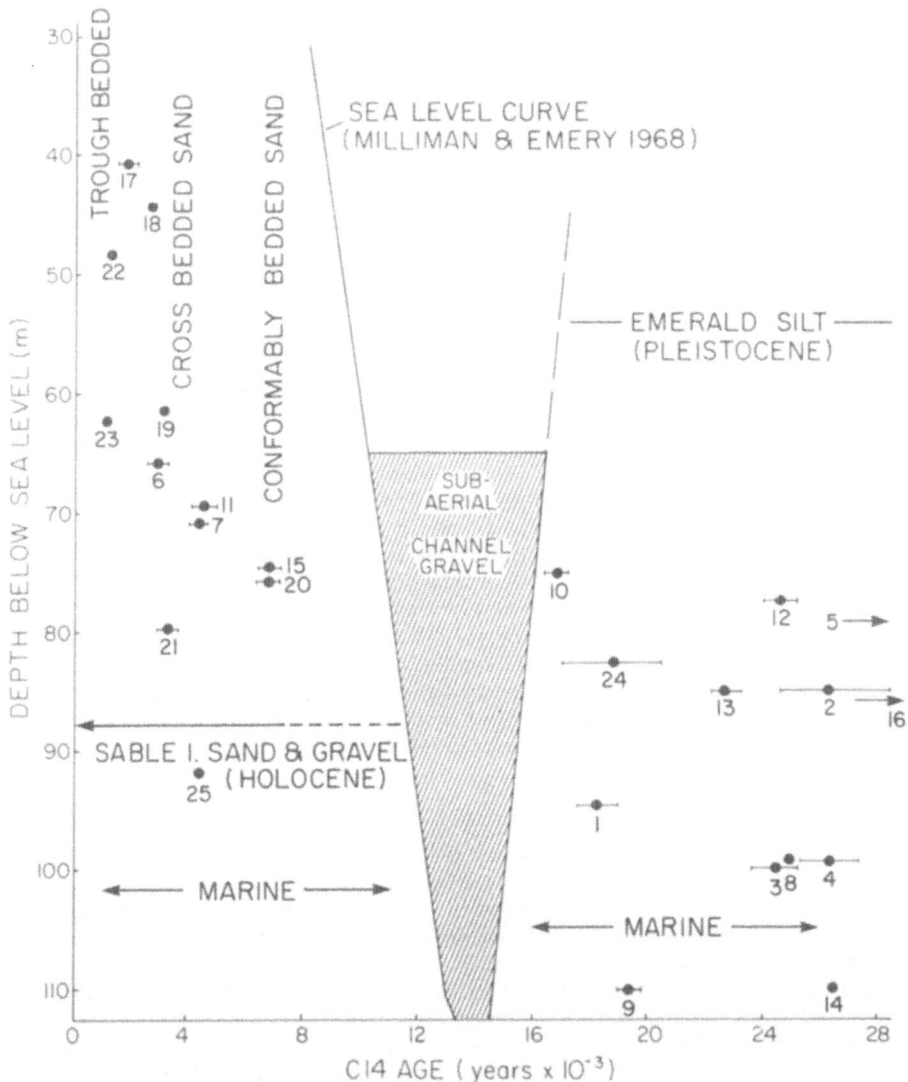

15. A plot of C 14 age against depth below sea level of 25 carbonaceous samples from boreholes on Banquereau (Amos and Knoll, 1987). This data set brackets the age of the transgression across the outer shelf banks and helps to define the environmental history from the Mid Wisconsinan to the Holocene. The transgression must be younger than the most recent glaciomarine date. The absence of dates between 8 ka BP and 16 ka BP corresponds to a period of regression followed by transgression. The sea-level curve of Milliman and Emery (1968) is shown for comparison.

indicate maximum coastal submergence at 13.5 ka BP up to 70m above present, followed by a rapid regression to a present depth of 65m (Belknap et al., 1986). Additional evidence for late glacial ice and associated crustal depression is found in the Truxton Swell area of the central Gulf of Maine where a Wisconsinan glacial surge at 17 ka BP (King and Fader,1986) deposited a widespread till over glaciomarine sediments. Late glacial ice on the land areas of the northern Fundy region resulted in significant rebound of the crust during the marine transgression of the Scotian Shelf where rebound was more or less complete by this time. Following recession of the ice, late glacial rebound raised the low sea-level position to its present depth of approximately 37m.

This example demonstrates that particle shape, sediment textural distributions and lithology, combined with analyses of seismic reflection data, can provide consistent lines of evidence for determining the former position of sea level and that these characteristics can be sensitive to changes in isostasy.

Chedabucto Bay

During a study of the surficial geology of Canso Bank and adjacent areas of the Scotian Shelf, (MacLean et al. 1977), glacial till was mapped in Chedabucto Bay, Cape Breton Island, in water depths as shallow as 70m. In the Strait of Canso engineering drill cores penetrated at least 29m of till. Preservation of glacial deposits at these shallow depths is dissimilar to other shallow areas of the shelf. As no tectonic or isostatic evidence exists from the area to account for the occurrence of the till in shallow water, it was suggested that late glacial ice may have covered the till and protected it from erosion during the transgression. It is also possible that the protected location of Chedabucto Bay from open ocean storms may have contributed to the preservation of till in these anomalous water depths. Other nearshore areas of the shelf, such as along the western shore of Nova Scotia and in Lunenburg and Mahone Bays (Piper et al.,1986), appear to be similar to the Chedabucto Bay setting and show the presence of late Wisconsinan ice and preserved till in shallow water. Thus the presence of glacial ice a short distance offshore, during the Late Wisconsinan marine transgression, may account for the nearshore preservation of till.

South Coast of Newfoundland

In the offshore region of the south coast of Newfoundland, Fader et al. (1982) found it difficult to determine the position of the low sea-level stand because the seabed is characterized by deep narrow depressions, bedrock ridges and steep rugged topography which make accurate sediment sampling difficult. However, samples were collected in transects on the flanks of the adjacent offshore banks such as Burgeo and St. Pierre where the topography is less severe. The textural data together with the distribution of till interpreted from the seismic reflection profiles suggests that the low sea-level position occurs in water depths of 110-120m. However, in the area to the east of the Burin Peninsula till occurs in water depths as shallow as 91m, which indicates a small amount of subsequent warping of the low sea-level stand or the presence of Late Wisconsinan ice cover. Studies of raised marine deposits along the southwest coast of Newfoundland (Brookes, 1977) indicate that these materials were deposited at approximately 14 ka BP at the same time the adjacent offshore banks were emergent. This indicates considerable differential rebound between the adjacent offshore banks and the southwestern coastal areas of Newfoundland, a situation similar to the coastal zone and the Gulf of Maine. There is little additional offshore evidence to constrain such a model for the south coast of Newfoundland. Brookes et al.,

(1985) investigated postglacial sea-level change along the southwest coast of Newfoundland and determined that a marine limit of 44m occurred at 13.6 ka BP which subsequently fell to a low level of -11 to -14m at 6 ka BP. They suggested that either thick late Wisconsinan ice over western Newfoundland or ice on Labrador to the northwest may have accounted for the crustal depression.

Grand Banks of Newfoundland

On the northern area of Grand Bank, in an area known as the Hibernia discovery region, the low sea-level position is interpreted to occur at a depth of approximately 100-110m (Fader and King, 1981; Fader et al., 1986). Low gradients of the shelf combined with seabed iceberg scouring make its precise determination difficult. For the adjacent inner shelf area in the Avalon Channel, the low sea-level stand appears to occur at a depth of 90m (Fader and Miller, 1986). The southern Tail of the Bank south of the Hibernia region, however, presents an unusual set of characteristics unlike other areas along the Canadian margin. A large deposit of muddy sand to sandy mud occurs in a 30km wide by 200km long zone extending from the Tail of the Bank to Whale Bank along the southwestern edge of Grand and Whale Banks. The sediment is up to 30m in thickness and occurs in water depths as shallow as 55m. If a postglacial low stand of sea level occurred below this depth, the muddy deposit would have been eroded in the subsequent transgressing sea, in a similar manner to sediments on the flanks of the banks of the Scotian Shelf. Its preservation in water depths as shallow as 55m suggests: 1) that the low sea-level stand in this area is less or 2) that the muddy deposit is a more recent sediment formed after the transgression had occurred. The implications for option 1 could be explained by the presence of a late glacial ice dome on the Tail of the Bank which depressed the area and delayed isostatic rebound. However, few constraints can be placed on these hypotheses as the age of the sediment is as yet unknown.

IMPLICATIONS AND DISCUSSION

These examples where the low sea-level stand appears to be anomalous or warped help to illustrate the sensitivity of the data in determining the low sea-level stand and associated differential warping. In some areas the variations observed in the depth of the position are in general agreement with the adjacent land based observational record of postglacial sea-level fluctuation. In other areas a gap exists between the offshore observations and the adjacent land areas suggesting that late ice on land may result in large amounts of differential rebound. This is further compounded by areas of very rough seabed morphology where the low sea-level stand is difficult to determine.

The occurrence of the silt-sized fraction immediately below the low sea-level stand is difficult to explain as a paleosea level indicator in light of the known depth of sediment transport and the wave/current regime at the present day shoreface along the Nova Scotia coast (Forbes, 1987). It would be expected that the silt content would not form a significant component of the sediment in water depths much less than 150m. It may be that the silt was deposited later as the transgression migrated to shallower water and it is coincidental that the present values for silt content increase in water depths at and beyond 110m.

The zone of eroded till surrounds the coastal areas of eastern Nova Scotia and extends offshore on the inner shelf to water depths of 110m, a distance of approximately 20km. One of the main criticisms of the model for the transgressive removal of till on the inner shelf is the idea that perhaps till was never deposited across this zone. The area is predominantly underlain by resistant Paleozoic metasediments and granites as shown on the sonogram of Figure 16, which may have supplied only a small amount of material to the glaciers which crossed the shelf. It has further been suggested that the till and other glaciomarine sediments only thicken over the Coastal Plain sediments from which material could easily be eroded. Figure 17 shows the distribution of the major moraines exposed at the seabed on the Scotian Shelf and the zero edge of the Coastal Plain section which thickens rapidly toward the shelf edge. From this distribution it can be seen that many of the moraines, some of which are up to 50m in height, overlie Paleozoic bedrock north of the Coastal Plain section and that their origin is not dependent on the occurrence of easily eroded Coastal Plain source rocks. This inner shelf absence of till, in a narrow zone between the continuous till to the south across the shelf and the till-covered Nova Scotian mainland, is therefore attributed to erosion during the late Wisconsinan-Holocene transgression.

With the exception of the areas noted above, the position of the low sea-level stand of 110-120m appears to be regionally consistent across the entire continental shelf of southeastern Canada. From this observation we can draw several conclusions concerning the isostatic and glacial history of the area. The consistent depth of the terrace suggests that very little warping has occurred along the shelf since its formation or that warping has occurred and the crust has returned to its former position. Where warping does occur it appears to be confined to the coastal areas. It also suggests that the ice had probably receded from the shelf before this time or else the terrace would not be as continuous or would be locally warped. The complex structural and stratigraphic relationships seen in coastal Maine and in the nearshore of the western Scotian Shelf, where glacial ice and the sea were in close proximity, do not occur in association with the low sea-level position on the central and outer shelf. This further supports the idea that the ice had retreated before the formation of the lowstand. The distribution of modified iceberg furrows adjacent to the low sea-level stand also suggests that the scours were formed before sea level reached its lowest position and that glaciers had receded from the shelf before this time.

Quinlan and Beaumont (1982) reconstructed the morphology of Late Wisconsinan ice over the Atlantic region from the postglacial relative sea-level record. They calculated a maximum lowering of relative sea level of 70m which is considerably less than the 110-120m lowering suggested by the geological evidence. They assumed that the time of maximum ice cover was 18 ka BP and further suggested that no obvious way to determine the rate of ice retreat was possible. Subsequent to their study, King and Fader (1986) proposed a model and a chronology for the Wisconsinan glaciation and deglaciation of the southeastern Canadian offshore which indicated maximum glaciation during the early-middle Wisconsinan. The detailed glacial history in King and Fader (1986), based on a study of seismic reflection and sample data, together with a conceptual model of glacial deposition, indicates that crustal rebound may have begun much earlier and that by 18 ka BP the offshore would have rebounded substantially from the effects of the ice load, including the migration of the glacial forebulge. As further suggested by Quinlan and Beaumont (1982), their ice model reconstruction now appears to have been merely representative of the later stages of recession of a more extensive and thicker ice sheet and this shift in the timing of maximum glaciation may account for some of the differences in the modelled and observed late Wisconsinan low sea-level stand in the southeastern Canadian offshore. Given the amplitude (15-20m) of the shoreface in the present analogous situation of the Nova Scotian eastern shore (Forbes, 1987), it is possible to interpret the actual position of the low sea level

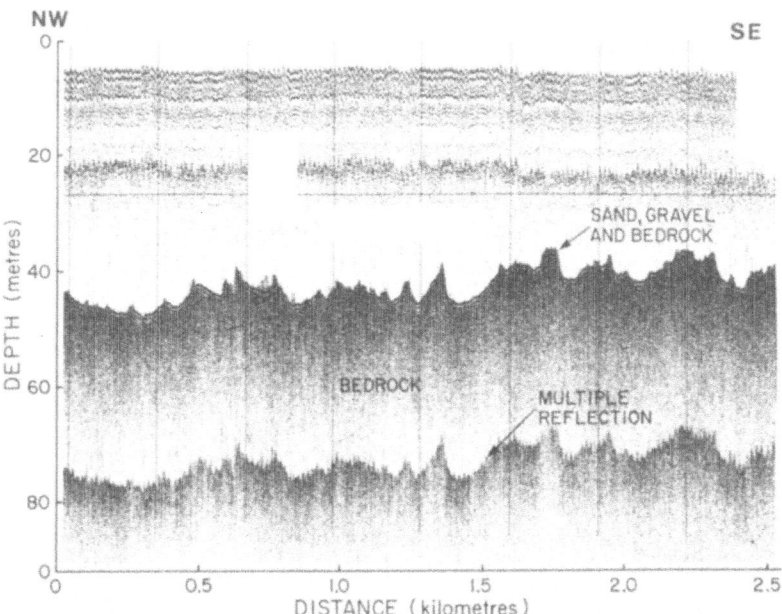

16. A Huntec DTS profile across the inner shelf off southwestern Nova Scotia, showing bedrock exposed at the seabed. This acoustic signature of a hard seabed with no penetration and subbottom reflections is characteristic and typical of many inner shelf areas above the low sea-level stand.

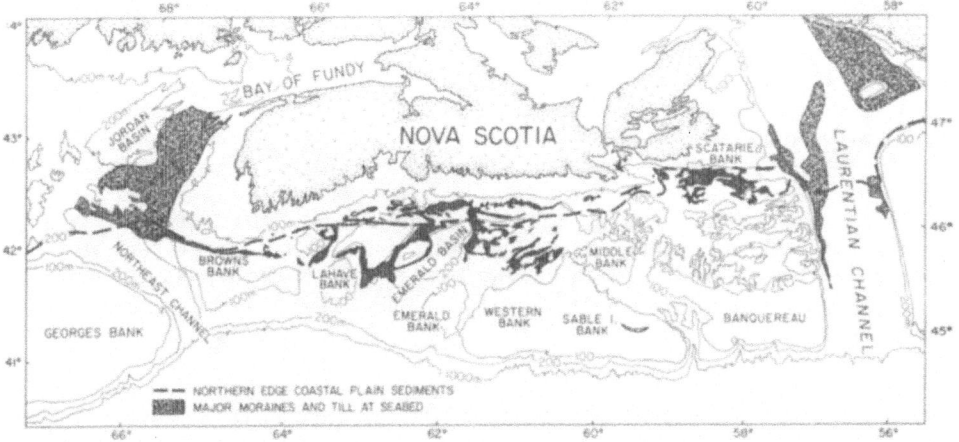

17. Map of the Scotian Shelf showing the zero edge of the semi-consolidated, southward thickening, Coastal Plain section and the major moraines exposed at the seabed, many of which are large ridges up to 50m in height. The largest moraines occur north of Emerald Basin overlying Paleozoic bedrock north of the Coastal Plain section. Moraines are not found on the inner shelf landward of the 100m contour. Their absence in this zone is attributed to erosion by the transgression.

to be 15-20m above the observed terraces. This interpretation, together with the effects of early-middle Wisconsinan isostatic recovery, lessens the disparity between the observed paleoshoreline features and geodynamically predicted relative sea level.

SUMMARY AND CONCLUSIONS

This paper summarises the history and evidence for the determination of a Late Wisconsinan low sea-level stand and a marine transgression for the southeastern Canadian offshore. A wide variety of data sources, including seabed samples and geophysical information, together with visual observations, identify a low sea-level stand that is often marked by a submarine terrace at a depth of 110-120m. Radiocarbon dates from sediments above and below the terrace indicate the age of formation at approximately 15 ka BP. The following are the major geological characteristics that provide information for an interpretation of the low sea-level stand and the implications of its occurrence.

1. Terraces at a depth of 110-120m are widespread across the shelf and are developed mainly in glacial sediments and to a lesser extent in bedrock.

2. The absence of silt and clay-sized sediments above the depth of occurrence of the low sea-level stand and their presence below is a useful indicator in defining the position of the low sea-level stand on a regional basis.

3. Well sorted sands and rounded to subrounded gravel occur above the terrace in contrast to angular gravel sized sediments below. Disk-shaped cobbles are common above 110-120m.

4. Unconformities across glacial sediments are widespread on areas of the shelf above the low sea-level stand and rare in the deeper basins. This results in the distribution of isolated remnants of glaciomarine sediments and till on the inner shelf and bank areas.

5. Till is generally absent on areas of the shelf above the terrace, particularily on the inner shelf where exposed bedrock dominates the seabed.

6. Iceberg furrows are rare and modern on the bank areas above the terrace; fresh and abundant in water depths below 160m; and eroded and infilled adjacent to and below the terrace.

7. The surface of the Tertiary succession underlying C 14 dated late Wisconsinan and Holocene deposits above the terrace on the Grand Banks of Newfoundland is leached and shows mineralogic evidence of soil formation and subaerial exposure in water depths of 70-100m.

8. The low sea-level stand occurs at approximately the same depth across the shelf except in some coastal areas where late glacial rebound of the adjacent land areas has resulted in the occurrence of the low sea-level stand in shallow water. This suggests that the ice had receded from the outer shelf much earlier and that rebound was largely complete before the low stand was developed. With the exception of the Tail of the Bank area of the Grand Banks of Newfoundland, it appears that little differential warping of the terrace on the outer shelf has occurred since its formation.

ACKNOWLEDGEMENTS

I thank the Officers and Crew of the CSS HUDSON and CSS DAWSON for their assistance in the collection of the data. The Technicians of the Program Support Subdivision of the Atlantic Geoscience Centre provided the technical expertise to collect high quality data. D. Clattenburg provided most of the sedimentological analyses of the seabed samples and R.O. Miller provided the field and laboratory support for the analyses of the data and the design of the diagrams. H. Joyce interpreted and compiled the distribution of boulders on the Grand Banks of Newfoundland. I also wish to thank D.B. Scott, Dalhousie University, for spirited discussions on sea levels, C.F.M. Lewis and C.L. Amos for ideas and suggestions, and in particular L.H. King for his insight and earlier work on the Scotian Shelf. The paper was reviewed and improved by C.F.M. Lewis, R.B. Taylor, D.B. Scott and D.F. Belknap.

REFERENCES

Amos, C.L. in press. Quaternary sediments and bedforms of the Scotian Shelf observed from the submersible Pisces IV. Spec. Publ. G.S.C.

Amos, C.L. and Knoll, R.G. 1987. The Quaternary sediments of Banquereau, Scotian Shelf. *Geological Society of America Bulletin*, v. **99**, p. 244-260.

Belknap, D.F., Shipp, R.C. and Kelley, J.T. 1986. Depositional setting and Quaternary stratigraphy of the Sheepscot Estuary, Maine: a preliminary report. *Géographil Physique et Quaternaire*, v. **6**, p. 55-69.

Boyd, R. and Penland, S. 1984. Shoreface translation and the Holocene stratigraphic record: examples from Nova Scotia, the Mississippi Delta, and eastern Australia. *Marine Geology*, v. **60**, p. 391-412.

Brookes, I.A. 1977. Radiocarbon age of Robinson's Head moraine, west Newfoundland, and its significance for postglacial sea-level changes. *Canadian Journal of Earth Sciences*, v. 14, p. 2101-2120.

Bruun, P. 1983. Beach scraping - is it damaging to beach stability? *Coastal Engineering*, v. **7**, p. 167-173.

Fader, G.B. in press. Submersible observations of iceberg furrows and sand ridges, Grand Banks of Newfoundland. G.S.C. Spec. Paper.

_____, King, L.H. and MacLean, B. 1977. Surficial geology of the eastern Gulf of Maine and Bay of Fundy. Geological Survey of Canada, Paper 76-17; Marine Sciences Paper 19, 23 p.

_____ and King, L.H. 1981. A reconnaissance study of the surficial geology of the Grand Banks of Newfoundland. Current Research Part A, Geological Survey of Canada, Paper 81-A, p. 45-81.

_____ and Miller, R.O. 1986. A reconnaissance study of the surficial and shallow bedrock geology of the southeastern Grand Banks of Newfoundland. In Current Research, Part B, Geological Survey of Canada Paper 86-1B, p. 591-604.

Forbes, D.L. 1987. Shoreline Sediment Distribution. In *Coastal Sediments* 87, v. 1, p. 694-709.

Grant, D.R. 1971. Glacial deposits, sea-level changes and pre-Wisconsin deposits in southwest Nova Scotia, Geological Survey of Canada Paper 71-B, p. 110-113.

Hutchins, R.W. 1974. Computer simulation of a transiently excited underwater sound projector. Ocean '74, Proc. in. Conf. on Engineering in the Ocean Environment, Halifax, Nova Scotia, 2, p. 115-119.

_____, McKeown, D.L. and King, L.H. 1976. A deep tow high resolution seismic system for continental shelf mapping. Geoscience Canada, v. **3**, p. 95-100.

James, N.P. and Stanley, D.J. 1968. Sable Island Bank of Nova Scotia: Sediment dispersal and recent history. *American Association of Petr. Geol. Bull.*, 52, p. 2208-2230.

King, C.A. 1972. *Beaches and Coasts*, Second Edition, Edward Arnold Publ., London, U.K., 570 p.

King, L.H. 1967. Use of conventional echo-sounder and textural analyses in delineating sedimentary facies - Scotian Shelf. *Canadian Journal of Earth Sciences*, v. 4, p. 691-708.

_____ 1970. Surficial geology of the Halifax - Sable Island map-area. Geological Survey of Canada Marine Science Paper 1. 16 pp.

_____ 1980. Aspects of regional surficial geology related to site investigation requirements - Eastern Canadian Shelf. In *Offshore Site Investigation*, D.A. Ardus (ed). Graham Trotman Ltd., London, 291 p.

_____ and Fader, G.B.J. 1986. Wisconsinan glaciation of the continental shelf - southeastern Atlantic Canada. Geological Survey of Canada Bulletin 363, 72 pp.

Letson, J.R.J. 1980. Sedimentology of southwestern Mahone Bay. Unpublished M.Sc. thesis, Dalhousie University, Halifax, N.S.

Lewis, C.F.M. et al. in press. Estimating iceberg scouring rates for the Grand Banks of Newfoundland, in POAC '87, Fairbanks, Alaska.

McKeown, D.L. 1975. Evaluation of the Huntec ('70) Hydrosonde Deep Tow Seismic System, B.I.O. Report BI-R-75-4.

Piper, D.J.W. et al. 1986. Quaternary Geology. In M.J. Keen and G.L. Williams (eds), Geology of Canada. G.S.C. Spec. Report.

Quinlan, G. and Beaumont, C. 1981. A comparison of observed and theoretical postglacial relative sea level in Atlantic Canada. *Canadian Journal of Earth Sciences*, v. **18**, p. 1146-1163.

_____ and _____ 1982. The deglaciation of Atlantic Canada as reconstructed from the postglacial relative sea-level record. *Canadian Journal of Earth Sciences*, v. **19**, p. 2232-2246.

Scott, D.B. and Medioli, F.S. 1982. Micropaleontological documentation for early Holocene fall of relative sea level on the Atlantic coast of Nova Scotia. *Geology*, v. **10**, p. 278-281.

Scott, D.B. and Greenburg, D.A. 1983. Relative sea-level rise and tidal development in the Fundy tidal system. *Canadian Journal of Earth Sciences*, v. **20**, p. 1554-1564.

Segall, M.P., Buckley, D.E., and Lewis, C.F.M. 1987. Clay mineral indicators of geological and geochemical subaerial modification of near-surface tertiary sediments on the Northeastern Grand Banks of Newfoundland. Geological Survey of Canada Contribution #40186, 31 p.

Sonnichsen, G., Fader, G.B.J., and Miller, R.O. 1987. Compilation of seabed sample data from the Scotian Shelf, the Western Grand Banks of Newfoundland and adjacent areas. G.S.C. Open File 1430.

Stanley, D.J., Drapeau, G. and Cok, A.E. 1968. Submerged terraces on the Nova Scotian Shelf. *Annals of Geomorphology*, v. **7**, p. 85-94.

HOLOCENE RELATIVE SEA-LEVEL CHANGES AND QUATERNARY GLACIAL EVENTS ON A CONTINENTAL SHELF EDGE: SABLE ISLAND BANK

D.B. Scott, R. Boyd, M. Douma and F.S. Medioli
Centre for Marine Geology
Dalhousie University
Halifax, N.S. B3H 3J5, Canada

S. Yuill
Jacques-McClelland Geosciences
1046 Barrington Street
Halifax, N.S. B3H 2R1, Canada

E. Leavitt
Mobil Oil Canada, Ltd.
P.O. Box 800
Calgary, Alberta T2P 2J7, Canada

C.F.M. Lewis
Atlantic Geoscience Centre
Bedford Institute of Oceanography
Dartmouth, Nova Scotia B2Y 4A2, Canada

ABSTRACT. Here we present results from the first continuous core that extends through the upper 150 m of section on Sable Island. This core was positioned in the centre of a 400 m deep buried channel that runs underneath Sable Island. We believe the channel to be analogous with subglacial tunnel valleys observed in northern Europe.

Peat layers from platform borings and a vibracore, just offshore, provide some of the deepest and oldest postglacial sea-level points ever recovered on the North American margin: from -48.8 m (11 kybp) to -35.3 m (8.1 kybp). Combining these points with previous sea-level data, a rate of relative sea-level (RSL) rise of 62 cm/100 yrs is calculated between 8.1 kybp and 4.5 kybp with a decrease of RSL rise to 30 cm/100 yrs from 4.5 kybp to the present. If the faster rate is extrapolated back to 15,000 ybp, we obtain a maximum sea-level lowering of about 78 m in the late Wisconsinan.

Geological Survey of Canada Contribution No. 24788.

D. B. Scott et al. (eds.), Late Quaternary Sea-Level Correlation and Applications, 105–119.
© 1989 by Kluwer Academic Publishers.

The Holocene section of sand is underlain by a series of glacial and glacial-marine clays and sands. Beneath these (at 65-80 m) is what appears to be an interglacial (or interstadial?) sand that is similar to the Holocene sand body. The lowest 60 m of the section is composed of thick, dense clay, which contains a glacial-marine microfossil assemblage. The sediment character and microfossil content are similar to the glacial-marine Emerald Silt observed in most intra-shelf basins inshore of Sable Island. Although we have no samples below 150 m, seismic reflectors in the 150-400 m section of the channel fill suggest chaotic cut and fill deposits, characteristic of the first stage of tunnel-valley filling. The 60 m clay unit may represent the last stage of channel fill (the "quiet phase") toward the end of the glacial interval. The position of the deposits above suggests that this major glacial cutting event occurred not during the isotope stage 2 world-wide Wisconsinan, but during an earlier (perhaps stage 4) glacial.

1. INTRODUCTION

The sea-bed topography of the continental shelf of eastern Canada is distinct from almost any other continental shelf, including that just to the south off New England. The inner shelf regions contain many overdeepened, isolated intra-shelf basins that are commonly 300-400 m deep (King and Fader 1986). The northern Scotian Shelf has a complex pattern of interconnected basins and they cut into Mesozoic and early Tertiary marine sedimentary rocks. High resolution seismic reflection profiles of the outer Scotian Shelf, and in particular those near Sable Island, indicate occurrences of buried channels with dimensions, and possibly origins, similar to those in the basins of the inner shelf.

Sable Island itself (Fig. 1) directly overlies a 400 m deep N-S trending channel, and is the only emergent feature on the outer continental shelf of eastern North America. The island therefore provides a unique platform for drilling. Previous shallow drilling on the island (<30 m, Scott et al. 1984a; Ruffman et al. 1985) and most of the geotechnical borings in adjacent shallow waters did not penetrate the channels due to insufficient drilling depth or inapropriate location. Although numerous exploration wells have been drilled in the area, little precise information exists on the location or extent of Quaternary sedimentation, because sampling did not commence until after the surface casing was set through the upper 400 m. The only well sampled continuously from the surface was Mobil C67 well on Sable Island, but cuttings taken at 3 m intervals were of low quality. In addition, the well was not drilled in a channel (Hardy 1974).

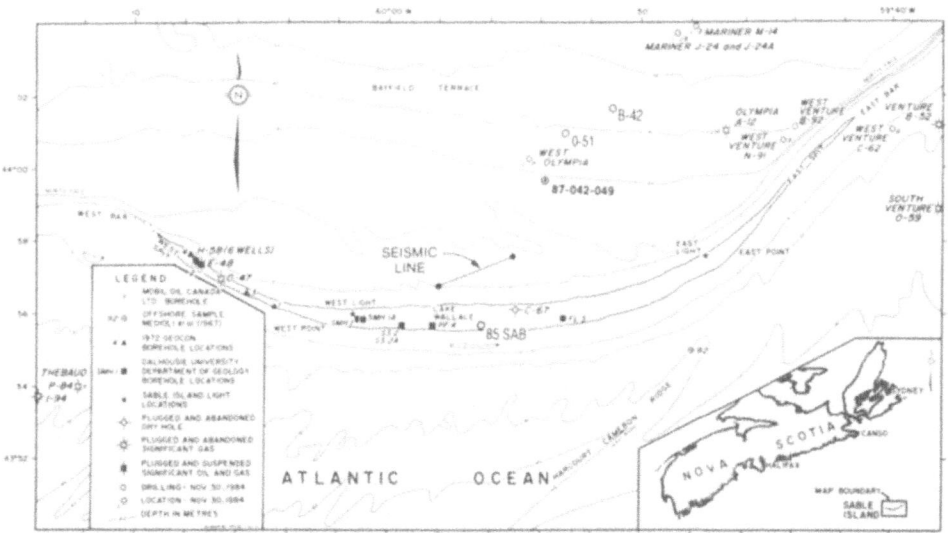

Fig. 1. Location of Sable Island with bathymetric contours and locations of all shallow boreholes around Sable Island up to 1987, including the SAB85 borehole and C67 locations. The boreholes 0-51, B-42, and vibracore 87-042-049 had buried peats in them. Map modified from Ruffman et al. 1985. The seismic line is the one shown in Fig. 3.

Scott et al. (1984a) primarily investigated Holocene sea-level changes in this area and were able to obtain data points as old as 7,000 ybp. However, there is considerable disagreement over the amplitude of total RSL rise that has taken place since the end of the Wisconsinan. King and Fader (1986) speculated that RSL was as much as 115 m lower than present at 15,000 ybp. Scott and Medioli (1982) suggested only about 30 m of RSL rise since the end of the last glaciation at the Nova Scotian coastline. Because Sable Island has been emergent since the end of the last glaciation, we hoped that a long core from this platform would provide us with a continuous record of postglacial RSL.

A consortium of university, industrial and government researchers obtained a long borehole sequence on Sable Island to interpret seismic profiles and to provide information on older sea levels. Because the borehole was located on the axis of a buried channel, the results of drilling provided the first evidence for the origin of the buried channels and key information for deciphering the glacial processes that may have influenced the Scotian Shelf.

2. METHODS

The borehole site at N43°55'33.7", W59°56'31.9" was located in the
axis of a major channel projected from evidence on multi-channel
seismic lines about 2 km offshore. The position of the borehole on
the island was determined to within 5 m using Argo high precision
positioning stations provided by McElhanney Offshore Surveys. The
seismic profiles used were obtained in a high-resolution marine
seismic survey, conducted for Mobil Oil Canada, which was originally
used to identify potential drilling hazards at the West Olympia site.
Seismic survey lines spaced 400 m apart traversed parallel to the
northern shoreline of Sable Island, with the closest line ranging 1-3
km from the island and 2-2.2 km from the C67 and SAB85 boreholes (Fig.
1). A minisleeve exploder source was used, recording twelvefold
coverage to a depth of two seconds. Wave equation migration
techniques were applied to lines near the borehole to obtain a
vertical resolution of 2 m. An Argo network used for position fixing
on the seismic survey was estimated to have an accuracy of better than
±10 m.

A fully equipped drilling installation consisting of a Longyear
38 drill, drilling platform, mud system, and geotechnical laboratory
was placed on the island to drill SAB85. Continuous sampling was
conducted throughout the entire section. Thin- and thick-walled
Shelby tubes permitted recovery of high quality geotechnical samples
in the unconsolidated material (sands and loose sandy silts) whereas
wire line HQ inner-tube core barrels were used to obtain full recovery
in the dense clay units.

All core samples were split, described, photographed, and tested
on site as required for the following geotechnical properties:
moisture content, coarse fraction (>74 um) for unconsolidated
sediments and shear strength as determined by torvane, hand
penetrometer and triaxial compression tests in consolidated clays
(Fig. 2). Portions of clay units were stored in foil and wax for
subsequent, more rigorous geotechnical testing. After initial
testing, the cores were sealed in plastic liners, wrapped, and placed
in D-tubes for shipment to Halifax. All cores were placed in cold
storage (4°C) at the Dalhousie core repository.

Gamma logging was performed on the upper 130 m of core (Figs. 2-
4). To identify sedimentary units, a standard facies logging method
was employed and supplemented by sieve grain size analyses and X-
radiographs. Benthonic foraminifera were used as paleoenvironmental
indicators using data from Medioli et al. (1986), Williamson (1983),
and Scott et al. (1980, 1984a, b) for comparison.

3. RESULTS

3.1 Seismic Stratigraphy

The minisleeve exploder profiles at the West Olympia site are high
quality, permitting clear recognition of a conformably bedded sequence

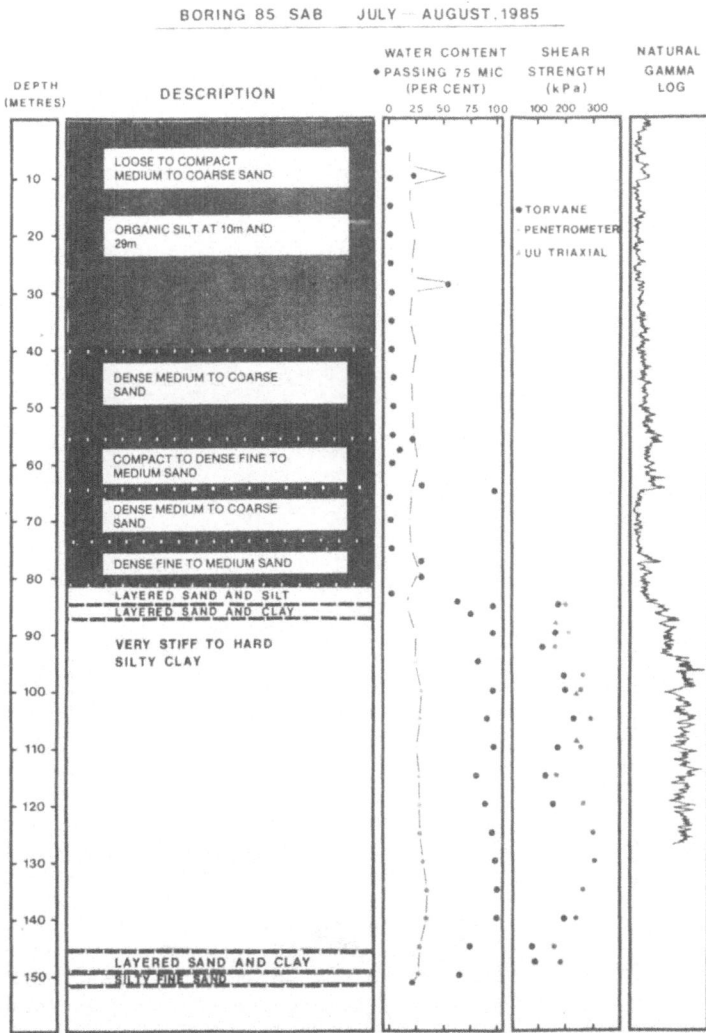

BOREHOLE RECORD

BORING 85 SAB   JULY — AUGUST.1985

Figure 2. Lithology of SAB85 borehole, also showing water content,
mud content, shear strengths, and gamma log.

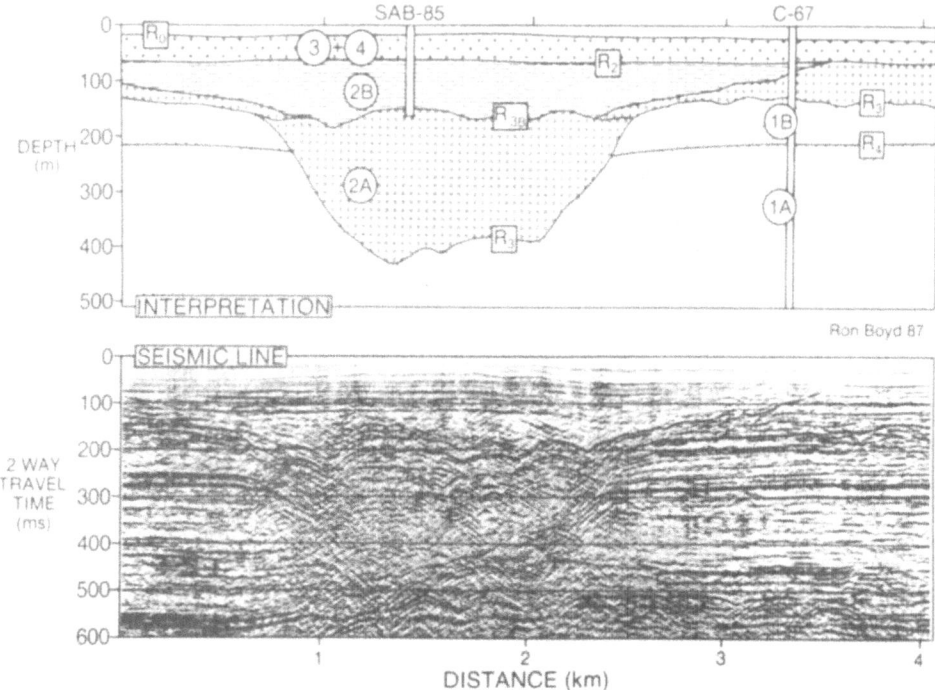

Fig. 3. Multi-channel seismic line (lower, run just north of SAB85 borehole) showing the deep buried channel. Upper part of figure is the interpretation with major reflectors (R numbers) and seismic packages indicated together with the position of the two boreholes discussed in this paper. SAB85 and C67 are projected onto the seismic line 2 km further south; this produces a variation in the depth of reflector R_4 between Figures 2 and 3 because of the regional dip of R_4.

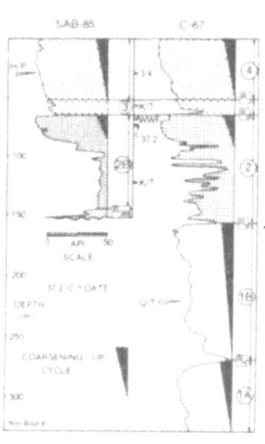

Fig. 4. Gamma logs of the SAB85 and C67 boreholes plotted against the major reflectors and seismic packages. H/P = Holocene/Pleistocene, Q/T = Quaternary/ Tertiary boundaries, K/T signifies Cretaceous/ Tertiary reworked fossils, WWF = warm water fauna, R numbers are reflectors, and circles numbers are packages-same patterns as Fig. 3. The lower 20 m of Gamma log from SAB85 is extrapolated, based on lithofacies. Although package 2 in C67 is equated with 2A and 2B in SAB85, the origin of package 2 outside the channel is unknown.

cut by a major network of deep channels (Fig. 3). Within the vicinity of the SAB85 and C67 boreholes, four seismic sequences can be recognized (Figs. 3 and 4) using the methods of Vail et al. (1977):

a) The oldest package, package 1, comprises a number of sequences separated by highly continuous, high-amplitude unconformities. Each sequence is composed of low- to moderate-amplitude, high-continuity reflectors that display parallel to divergent configuration and occasional shingled clinoforms. This style of seismic facies is characteristic of shallow continental shelf sedimentation and prograding-shelf phase deltas, consistent with Hardy's (1974) description of the Miocene Esperanto beds in the Tertiary Banquereau Formation.

b) Package 2 (which includes the main channel cut into package 1) has a linear asymmetric channel fill geometry and is up to 260 m thick, bounded by high-relief erosional unconformities. Within the channel fill, internal reflectors have low continuity, variable amplitude and a chaotic configuration. No coherent reflectors can be traced within package 2, but the seismic facies occasionally display a channel cut-and-fill pattern. Parallel reflectors of package 1 appear to be pulled up underneath the channel fill, indicating relatively high velocity in package 2 sediments. Within the channel (where SAB85 is located), two sequences of package 2 can be recognized: 2A with strong but chaotic reflectors and 2B which has a weak, almost transparent character. Sequence 2B is separated from 2 outside the channel by reflector R_{3B}.

c) Package 3 (capping the channel fill of package 2) also exhibits a channel fill geometry with sediments up to 11 m thick but it is much broader than package 2. The basal boundary of package 3 is a high-relief erosional unconformity, whereas the upper boundary is a flat surface, occurring about 73 m below sea level. The upper part of package 3 displays a small-scale channel cut-and-fill pattern near the Sable Island boreholes. The lower section is seismically opaque, having few coherent reflectors. Package 1 reflectors are pulled down under package 3 indicating relatively low seismic velocity in the upper channel fill.

Analysis of seismic data suggests that the main channel under Sable Island is asymmetric in cross-section with a maximum channel-axis depth of 408 m below sea level (all depth conversions assume an average sediment velocity of 1600 m/sec). The steeper western channel walls have average slopes of 30-35 degrees. The narrow central section of the main channel, occupied by sequence 2A, is only 1800 m wide whereas the channel fill of package 3 is 3.8 km wide. The channel has a thalweg of variable depth, which shows no consistent trend of deepening in any direction. Relief along the channel is 57 m over 2 km and several tributary valleys have base levels 60 m shallower than the main channel.

3.2 Lithostratigraphy

Determination of lithofacies in SAB85 was based on sedimentology, geophysical logging and geotechnical properties. The SAB85 borehole was cored continuously, with more than 90% recovery. In addition to information derived from the standard logging methods for sediment facies (e.g. Walker 1985), the data produced in SAB85 included mud (silt+clay) content, sediment strength and a log of natural gamma radiation (Fig. 2 and Jacques-McClelland 1985). In contrast, the Mobil C67 exploration well provided only rotary cuttings at 3 m intervals and the lithofacies determination was based on sediment composition, gamma and electric log data.

3.2.1 Lithofacies in SAB85. Three major facies can be recognized:
     1) Facies 1 occurs from 0 to 56.3 m and 64 to 77 m and consists of fine to coarse grained, fawn to gray sand, interbedded with thin mud, gravel and organic horizons. The sand shows evidence of erosional truncation, planar cross stratification, and parallel lamination. Grain size changes within Facies 1 indicate at least two major fining-upward trends (from 77 m to 64 m and 56.3 m to 29 m), both of which contain numerous minor internal fining-upward cycles. The sediment is primarily a mature quartz sand with minor (usually <10%) components of shell, granules, organic matter and mud.
     2) Facies 2 is interbedded sand, silt and clay. Alternating sands, silts and clays occur in Facies 2 at 56 to 64 m and 78 to 104 m and 143 to 151.5 m below the surface. The upper facies boundary was picked where mud content began to increase significantly and exceeded 10% below 78 m. The other boundaries mark either the beginning or end of thick, structureless clay sequences. Facies 2 comprises an overall coarsening-upward grain size trend whereas individual units begin with scoured erosional bases and show both coarsening- and fining-upward trends of sediments ranging from clay to gravel. Minor sediment components include shells and mud clasts. Several intervals within Facies 2 exhibit major sediment deformation structures including slumping, faulting, shearing, and fluid escape with individual beds dipping at 50-90°. Variations in geotechnical properties in Facies 2 parallel grainsize variability, and show relatively low undrained shear strengths of 90-250 kPa (Fig. 2).
     3) Facies 3 is primarily a very stiff, dark olive gray, silty clay at 104-143 m below the surface. These clays generally appear structureless even when examined by X-radiography and show little evidence of bioturbation. A thin lens of this clay occurs at 64.5 m, lying within Facies 1 (high mud content point at 65 m in Fig. 2). Minor components of Facies 3 are frequent wispy laminations and irregular patches of sand and silt, sediment deformation, loading structures, and erosional scours. Geotechnical parameters suggest that Facies 3 is normally consolidated to slightly preconsolidated with undrained shear strengths in the range of 180-300 kPa and water contents of 20-40% (Fig. 2). The silty clay is of medium plasticity, with average liquid and plastic limits of 45 and 21 respectively.

3.2.2 Lithofacies of C67. Three general lithofacies can also be recognized in the C67 borehole down to a depth that is equivalent to the channel base identified from seismic records.
     1) Facies 1 is interbedded sandstone and claystone. The

barren of microfossils, except for some shelly horizons. One level (67.66 m) contains a foraminiferal assemblage containing several formae of <u>Elphidium</u> <u>excavatum</u> (Miller et al. 1982), which suggests conditions that are warmer-than-present-day. The presence of this assemblage suggests that the 65-80 m sandy unit may represent a nonglacial interval. An accelerator C^{14} date from a shell at 71.92 m (37,000 ± 400 ybp, RIDDL#639) supports a correlation of this interval with the oxygen isotope stage 3 interstadial.

At 80.38 m, benthonic foraminifera again appear in association with deposits that probably represent the end of a glacial period. Abundant specimens are first observed at 86.80 m where common <u>Elphidium</u> species appear to represent marine ice-front conditions (some reworked T-K forms are also present).

In the clay unit itself, foraminiferal assemblages, similar to the lower part of the glacial-marine Emerald Silt (Vilks and Rashid 1976; Scott et al. 1984b; Scott and Medioli, in press;), are observed: <u>Elphidium</u> and <u>Cassidulina</u> together with high percentages of T-K reworked forms. In the lower part of the clay unit (137-140 m), there is a barren zone, perhaps denoting a high sedimentation rate that diluted the microfossils. In the 140-150 m level, more diverse microfaunas occur indicating that an earlier period of glacial conditions may have just been starting. There are relatively few T-K reworked forms in the 149-151 m section, possibly because of reduced glacial erosive activity in the first stages of a glaciation.

3.3.2   <u>C67 Borehole</u>.   The C67 sequence provides a non-channel reference section. Detailed comparison with SAB85 is difficult, however, because C67 samples were taken only every 3 m and are rotary cuttings. However, gross stratigraphic features can be determined. In the upper 100 m, most samples are completely barren of foraminifera. At about 100 m foraminifera begin to appear, mainly as glacial-like assemblages, although the numbers are so low that no definite paleoenvironment can be assigned. At about 170 m (where the mud content increases), more foraminifera are present and the first reworked T-K forms occur, suggesting glacial origins for these deposits. At about 220 m total numbers of foraminifera increase substantially and the assemblage takes on a warm water aspect probably of Mio-Pleistocene character, with reworked T-K forms decreasing markedly.

Hardy (1974) placed the Plio-Pleistocene boundary at approximately 260 m in the C67 borehole. Re-examination of the samples in the upper 600 m of this borehole (K. MacKinnon and F. Gradstein, Dalhousie) indicate that the boundary is at 220 m based on the first occurrence of <u>Asterigerina</u> <u>guruchi</u> (Gradstein and Agterberg 1982). This boundary sequence has been eroded away at SAB85 and is considerably higher than the base of the channel at 400 m.

3.3.3   <u>New Sea-level Points</u>.   The SAB85 borehole recovered no new sea-level points. One organic horizon at 44 m proved to have insufficient carbon even for an accelerator date. The other organic layer at 28.5 m was a lagoonal deposit, which is not a good sea-level indicator.

lithology at 177 to 408 m below the surface is fine to coarse gray
sandstone interbedded with brown claystone and minor components of
mudstone, lignite, and magnetite. Spontaneous potential log traces of
Facies 1 suggest relatively constant porosity and permeability (and
hence inferred texture), with the exception of a tighter, more
impermeable unit from 258 to 281 m below sea level.

2) Facies 2 is a slightly muddy sandstone. Between 98 and 177 m
the lithology is primarily fine to coarse gray sandstone with minor
red claystone stringers, shell fragments, plant fragments, and an
occasional muddy matrix. The electric logs show a highly variable
response in Facies 2, with an overall trend of decreasing porosity and
permeability.

3) Facies 3 is quartz sandstone. This facies is a distinct celom
to white, mature, quartz sandstone occurring at 0 to 177 m below the
surface. It contains a low mud content and occasional shell (Ostrea
sp.) fragments. Electric log response in Facies 3 becomes less
variable and shows an alternating upward increase/decrease in porosity
and permeability.

3.3   Benthonic Foraminifera and C^{14} Dates

3.3.1 SAB85. Fifty levels were examined qualitatively for benthonic
foraminiferal assemblages. The majority of samples from the upper 80
m were barren of foraminifera, although an organic layer at 28.52 m
contained an abundant benthonic fauna composed of Elphidium spp.,
indicating high salinity,, open lagoon conditions and possibly
relatively deep water (>10 m). Also present were some planktonic
forms indicating open exchange with the ocean. A C^{14} date (3,365 ± 70
ybp, Lab #TO-209) places this layer as a late Holocene lagoonal
interval, consistent with other organic deposits found at similar
depths (Scott et al. 1984a). This date initially seemed very young in
comparison with our sea-level dates at 10-15 m subsurface that are
older (Scott et al. 1984a). However, the microfauna indicate that
this deposit represents a lagoonal environment with water depths of 15
to 20 m, and therefore does not contradict a sea level of about -10 m
for that time period. These relationships emphasize the need to
obtain micropaleontological information for organic deposits before
using them as sea-level indicators.

The only other significant occurrence of microfossils in the
upper 80 m is the zone between 56 and 65 m. At 64.45-64.68 m, a fauna
characteristic of late Pleistocene glacial-marine conditions is found
together with large numbers of reworked Tertiary-Cretaceous (T-K)
forms. The reworked T-K forms are also found in a sample at 56.30 m,
indicating glacial-marine conditions at least up to that level. A
second sample, just below (57.77 m), has an abundant Elphidium-
Cassidulina warm ice-margin assemblage with moderate numbers of T-K
forms. Reworked T-K forms only occur in the Quaternary during glacial
erosive intervals when ice action places in suspension large amounts
of Tertiary sediment from the shallow bank areas landward of Sable
Island Bank (Scott and Medioli, in press).

Below the clay layer, at 65 m, the coarse sands are generally

However five new points are included here from other sources that add to the sea-level history (Fig. 5): 1) -48.8 m comes from an offshore platform boring near the eastern tip of Sable Island (B-42 on Fig. 1), which recovered a peat horizon, C^{14} dated at 10,950 $\pm$ 70 ybp (lab #TO-184); 2) from another offshore borehole (0-51, Fig. 1), at -40.8 m was dated at 9930 $\pm$ 225 ybp (lab #GX-13485); 3) reworked peat, recovered from a shallow borehole done on the island itself in 1983 by Scott,-21.4 m dated at 8525 $\pm$ 250 ybp (lab #GX-10386); 4) peat from a vibracore (87-042-049, Fig. 1) near the island, at -35.6 m, dated at 8445 $\pm$ 125 ybp (lab #GX-13971); 5) another freshwater peat from the same core at -35.35 m dated at 8100 $\pm$ 235 ybp (lab #GX-13970). Of these points, only point 5 can be related accurately to a former sea level since it contains thecamoebians indicating that it formed at sea level in the Sable Island sand body. The remaining additional points are freshwater peat layers that formed an indeterminate amount above the higher tide mark. The curve in Figure 5 is drawn through the points which are vertically well constrained in relation to former sea levels, i.e. the salt marsh dates from previous work and point 5 here. Two distinct trends are observed: an earlier, more rapid rate of RSL between 8.1 kybp and 4.5 kybp (62 cm/100 yrs) and a later slowing of RSL (30 cm/100 yrs, 4.5 kybp to present). If the highest rate is extrapolated from 8.1 kybp to 15 kybp (probable age of the last low stand, Fader and King, 1986; Amos and Knoll, 1987), we obtain a lowest sea level since the Wisconsinan glaciation of -78 m below present. The difference in elevation between point 3 and points 4 and 5 provides a sense of what the relief of the island was prior to the Holocene transgression; about 15-20 m or similar to the present day dune system.

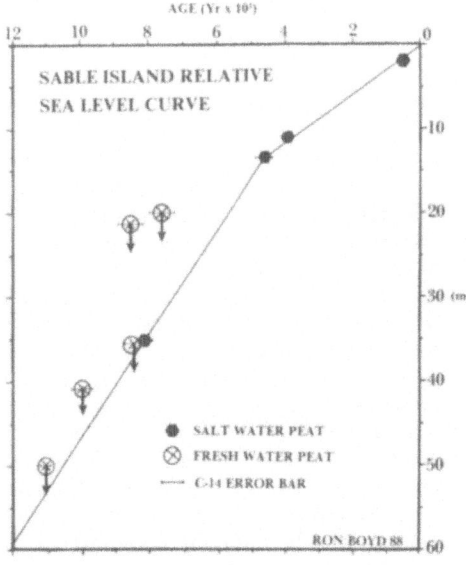

Fig. 5. New sea-level curve from Sable Island Bank. Arrows indicate that these points are minimum RSL lowering, i.e. sea level could have been lower than these levels.

4. DISCUSSION

The five new sea-level points added to the Sable Island curve constitute a suite of the oldest reliable sea-level points yet obtained on the edge of the Scotian Shelf. To the best of our knowledge, the -48.8 m peat is the deepest and oldest postglacial peat ever recovered on the east coast of Canada. The maximum lowering of sea level that we project (-78) is less than the 100-115 m sea level assigned by Fader (this volume) to the erosional 115 m terrace on Sable Island Bank for the same period. The extrapolated level agrees well with a seismic unit interpreted as the transgressive Holocene sand which begins in 65-80 m on the outer edge of Sable Island Bank (Scott et al. 1987). However, the position of the lowest sea level during the maximum late Wisconsinan remains a question because there are no dated surfaces older than 11 kybp. The 115 m level is 80 m deeper than the maximum lowering suggested by Scott and Medioli (1982) for the coastline, some 150 km north of this location. However, this depth difference can be explained by the dynamic position of the glacial forebulge which moved across the continental shelf following deglaciation (Quinlan and Beaumont 1981; Scott et al. 1987). The difference in maximum RSL rise between the two sites provides the first indication of what the amplitude of change might be between sea-level zones proposed by Quinlan and Beaumont (1981). For comparison, Quinlan and Beaumont (1981) predict a maximum lowering of about 60 m for the outer Scotian Shelf (Sable Island) and 30 m for the inner Scotian Shelf, the same as that measured by Scott and Medioli (1982) for Lunenburg Bay on the coast. Also Quinlan and Beaumont's (1981) model predicts sea-level fall in early postglacial times, not the rapid rise produced by simply extending our curve past 8.1 kybp.

Biostratigraphic evidence from the C67 borehole shows that the Quaternary/Tertiary boundary lies within a conformably bedded sequence at 220 m below present sea level. Seismic lines suggest an average dip of 1-2° to the SE for this sequence, thereby indicating the presence of a thickening wedge of Quaternary sediment south from Sable Island to the shelf edge. The position of the boundary lies within a coarsening upward sequence that is characterised by parallel to divergent seismic reflectors. These features suggest continued deposition across the Q/T boundary in a prograding shelf setting. An increase in reworked T-K microfauna above the boundary is the signal that indicates the onset of Pleistocene glaciations.

A major erosional event in the Quaternary history of the Sable Island region formed a widespread discontinuity herein termed reflector R_3. This event appears to have been regional in nature, but there are locally scoured, overdeepened valleys such as that in which the SAB85 borehole was drilled. This valley connects to a network of similar channels mapped from multi-channel seismic lines (Boyd et al. 1988). The channels all have bases deeper than 300 m but show no consistent trend of deepening along channel. The Tertiary and Quaternary shelf edges are located over 30 km south of Sable Island suggesting that the channels are not canyon heads deepening toward the continental slope. Similarly the channels are not fluvial in origin as they have a Quaternary age and no evidence exists for lowering of sea level below -200 m during the Quaternary.

The channels most likely formed from sub-ice meltwater streams confined within tunnel valleys (Boyd et al. 1988). In stratigraphic characteristics, channel cross-section, and drainage network, the Sable Island Bank tunnel valleys closely resemble those described on the north German coast by Ehlers (1981) and Kuster and Meyer (1979). Similar, but less well documented, valleys are found in the marine environment in the North Sea (Jansen 1976).

The majority of seismic package 2 is suggested to represent Pleistocene glacial deposition. Its lithological and log characteristics indicate wide variabiity and a complex history, dominated by channel erosion and subsequent infilling. The fine grained clay unit of lithofacies 3 in the SAB85 borehole occurs as a "plug" above the main tunnel valley. Microfauna indicate a glaciomarine, warm ice-margin environment for this unit. It may have formed in a fjord-like or estuarine setting protected by surrounding ice or by topography, if sea level was much lower (i.e. >70 m below present) or it may represent channelised reworking.

Seismic package 3 and 4 are primarily composed of sandy sediments in lithofacies 1. Microfauna in the SAB85 borehole indicate that the glacial marine environment of lithofacies 2 and 3, marked at its base by a thin clay layer at 65 m, changes to interstadial below 65 m (65-75 m below surface). The Sable Island sand body, a coarsening upward sequence of sand with its base at 56 m, is generally barren of microfossils but does have random beds of fine grained organic units which appear to be estuarine or lagoonal above 28.5 m (Scott et al. 1984a). This sequence, above 28.5 m, consistently provides C^{14} dates of less than 12,000 ybp on all types of organic remains. Because there is no present source of this sand, it has probably been derived during and since the Holocene transgression from underlying glacial sediments. This suggests that the glacial material below the sand is probably Wisconsinan in age and that the thin clay unit at 65 m below the surface may be a small erosional remnant of an earlier thick glacial marine clay. Carbon-14 dates of 30-40 kybp in the sands at 65-80 m suggest that the unit may be equivalent to the oxygen isotope stage 3 interstadial. If that is the case, and if there are no major unconformities between 65 and 90 m, then the thick clay unit and the majority of the glacial marine sediment in the channel drilled by SAB85, may correspond to the isotope stage 4 glaciation. Global isotope curves suggest that the strongest glacial event in the last 100,000 years is the stage 2 event, not stage 4. On this margin, stage 4 may have had a much greater impact which suggests that this part of eastern Canada may have been out of synchronization with the global signal.

5. ACKNOWLEDGEMENTS

Virtually everyone in the Centre for Marine Geology was involved in the logistics of this programme as it was being carried out on Sable

Island. However, J. Barrett, T. Duffett, and V. Baki made many trips to the airport with vital supplies that had to be flown out to us. A. Giddy and H. Brown were our cooks and kept our morale high. Drillers were F. Logan, R. Murphy, J. Whelan, D. Fullerton, M. Woodin, B. Hiltz, and R. Prosser. Project engineers were P. Green, J. Brown, S. Yuill, C. Silliphant, C. Yates, G. MacNeil, G. Crouse, and B. Van Lingan. Mobil Oil Canada provided us with substantial financial assistance as well as access to confidential information that helped us position the drill hole. We also thank all the service companies that were associated with this project who went beyond the call of duty to assure its success: Logan Drilling, Sea-Land Helicopters, Balder Ships, IMP Aviation, Maritime Tel & Tel, McElhanney Offshore Surveys and McKenzie Foods. Two government agencies assisted in our logistics on and off the island: the Canadian Coast Guard and Atmospheric Environment Services. Major funding was provided by a Natural Sciences and Engineering Research Council (NSERC) University-Industry grant to Scott, Boyd, Medioli, Jacques-McClelland Geosciences and Mobil Oil. Substantial funding also came from Mobil Oil and the Venture partners as well as ship and helicopter time. Additional funding was provided by the Canada Panel on Energy Research Development (PERD). NSERC operating grants to Scott, Medioli and Boyd funded subsequent research on the borehole material.

REFERENCES

Amos, C.L. and Knoll, R.G. 1987. 'The Quaternary sediments of Banquereau, Scotian Shelf'. Geological Society of America, Bulletin, 99:244-260.

Boyd, R., Scott, D.B., and Douma, M. 1988. 'Glacial tunnel valleys and Quaternary history of the outer Scotian Shelf'. Nature, 333(6168):61-64.

Ehlers, J. 1981. 'Some aspects of glacial erosion and deposition in North Germany', Annals of Glaciology, 2:143-146.

Gradstein, F.M. and Agterberg, F.B. 1982. 'Models of Cenozoic foraminiferal stratigraphy-Northwestern Atlantic margin', in Cubitt, J.M. and Reyment, R.A., eds. Quantitative Stratigraphic Correlation, John Wiley and Sons, Ltd., 119-170.

Hardy, I.A. 1974. 'Lithostratigraphy of the Banquereau formation on the Scotian Shelf: offshore eastern Canada'. Geological Survey of Canada, Paper 74-30, 2:163-174.

Jacques-McClelland Geosciences, Inc. 1985. '1985 Sable borehole project, report no. G042: report to the Centre for Marine Geology and Atlantic Geoscience Centre', 32 p.

Jansen, J.H.F. 1976. 'Late Pleistocene and Holocene history of the northern North Sea, based on acoustic reflection records'. Netherlands Journal of Sea Research, 10(1):1-43.

King, L.H. and Fader, G.B. 1986. 'Wisconsinan glaciation of the Atlantic continental shelf off southeast Canada'. Geological Survey of Canada, Bulletin, 363:76.

Kuster, H. and Meyer, K.-D. 1979. 'Glaziare Rinnen im Mittleren und nordostlichen Niedersachen'. Eiszeitalter und Gegenwart, 29:135-156.

Medioli, F.S., Schafer, C.T. and Scott, D.B. 1986. 'Distribution of recent benthonic foraminifera near Sable Island, Nova Scotia'. Canadian Journal of Earth Sciences, 23(7):985-1000.

Miller, A.A.L., Scott, D.B. and Medioli, F.S. 1982. 'Elphidium excavatum (Terquem): ecophenotypic vs. subspecific variation'. Journal of Foraminiferal Research, 12(2):116-144.

Quinlan, G. and Beaumont, C. 1981. 'A comparison of observed and theoretical postglacial relative sealevel in Atlantic Canada'. Canadian Journal of Earth Sciences, 6:1146-1163.

Ruffman, A., Miller, A.A.L. and Scott, D.B. 1985. 'Holocene rise of relative sea level at Sable Island, Nova Scotia, Canada: correction and note'. Geology, 13:661-663.

Scott, D.B. and Medioli, F.S. 1982. 'Micropaleontological documentation for early Holocene fall of relative sea level on the Atlantic coast of Nova Scotia'. Geology, 10:278-281.

Scott, D.B., Schafer, C.T. and Medioli, F.S. 1980. 'Eastern Canadian estuarine foraminifera: a framework for comparison'. Journal of Foraminiferal Research, 10(3):205-234.

Scott, D.B., Medioli, F.S. and Duffett, T.E. 1984a. 'Holocene rise of relative sea level at Sable Island, Nova Scotia, Canada'. Geology, 12:173-176.

Scott, D.B., Mudie, P.J., Vilks, G. and Younger, C.D. 1984b. 'Latest Pleistocene-Holocene paleoceanographic trends on the continental margin of Eastern Canada: foraminiferal, dinoflagellate, and pollen evidence'. Marine Micropaleontology, 9:181-218.

Scott, D.B., Boyd, R. and Medioli, F.S. 1987. 'Relative sea-level changes in Atlantic Canada: observed level and sedimentological changes vs. theoretical models'. Society of Economic Paleontologists and Mineralogists, Publication No. 41, 81-90.

Scott, D.B. and Medioli, F.S. In Press. 'Tertiary-Cretaceous reworked microfossils in Pleistocene glacial-marine sediments: an index to glacial activity'. In press to Marine Geology.

Vail, P.R., Mitchum, R.M. Jr., Todd, R.G., Widmier, J.M., Thompson, S.III, Sangree, J.B., Bubb, J.N., and Hatlelid, W.G. 1977. 'Seismic stratigraphy and global changes of sea level', in Payton, C.E., ed., Seismic Stratigraphy - Application of Hydrocarbon Exploration, American Association of Petroleum Geologists, Memoir 26, 49-212.

Vilks, G. and Rashid, M.A. 1976. 'Postglacial paleoceanography of Emerald Basin, Nova Scotia'. Canadian Journal of Earth Sciences, 13(9):1256-1267.

Williamson, M.A. 1983. 'Benthic foraminiferal assemblages on the continental margin off Nova Scotia: a multivariate approach'. Ph.D. Thesis, Dalhousie University, Halifax, Nova Scotia, 348 p.

Walker, R.G. (ed.). 1985. Facies Models. Geoscience Reprint Series 1, 317 p.

GEODYNAMIQUE DES LIGNES DE RIVAGE QUATERNAIRES DU CONTINENT AFRICAIN
ET APPLICATIONS

Pierre Giresse
Laboratoire de Recherches de Sédimentologie Marine
Université de Perpignan
66026 Perpignan, France

ABSTRACT. The world shorelines maps published in 1967 and 1981 show the
African continent with emerged strandlines dated at 2500-6000 years
B.P. and about 120.000 years B.P. Currently, we are striving for better
precision in the (recognition and) definition of Holocene shorelines;
in contrast, the significance of many Pleistocene terraces is being re-
assessed. Regions with several reproducible radiometric dates are relatively few: Morocco to northern Senegal, Angola to South Africa, Madagascar, Mascarene and Seychelles island group and some Red Sea coasts.
These coasts are the sites of positive epeirogenic movements of which
the rate and cause are more or less fully documented. The Atlantic
coast between southern Senegal and northern Angola does not have any
emerged dated Pleistocene marine shorelevels, and the Indian Ocean
coast is still relatively unexplored. In general, the pre-Holocene regression correlates with a dry period in the tropical and equatorial
latitudes, but corresponds to fluvial periods in the subtropical region as far south as Swakopmund in Namibia, and as far north as the latitude of Hoggar. The sequence of coastal environments is becoming
known with increasing precision in Senegal, Ivory Coast, Benin, Nigeria,
Gabon and Congo. Recent and fossil shorelines are special targets of
interest for mineral resource exploration because of their close relationship with both their mineral constituents and the physical forces
responsible for the accumulation of these minerals. Occurrences of
heavy minerals, associated with littoral features (shoreface, shore
and dune), are displayed on the West African coast of Gabon-Congo and
Senegal-Mauritania, on the shorelines of South-Africa and on the shelf
of Mozambique. Generally, human settlement coïncided with the proximity
of shorelines during high sea levels. Erosion and progradation of the
coast during the Late Holocene vary considerably along the coast; they
involve a combination of hydrologic, meteorologic and oceanographic
factors with the broad slow fluctuations of local eustatic character.

INTRODUCTION

L'état de nos connaissances sur les lignes de rivage quaternaires

D. B. Scott et al. (eds.), Late Quaternary Sea-Level Correlation and Applications, 121–152.
© *1989 by Kluwer Academic Publishers.*

du continent africain demeure, à ce jour, assez inégal. Certains secteurs privilégiés de l'Afrique de l'Ouest et de l'Afrique du Sud ont
été le site d'études répétées, parfois multidisciplinaires, qui permettent un bilan relativement précis qui aura valeur de référence pour
d'autres secteurs moins bien ou trop anciennement connus. Il demeure
qu'aujourd'hui encore, de longs traits de côte sont toujours peu ou pas
explorés, notamment en Afrique de l'Est.

Le document présenté repose en grande partie sur les rapports
d'activité de la Sous-Commission Afrique des Lignes de rivage de l'IN-
QUA de 1977 à 1987. A partir de 1983, le programme "Variations du niveau de la mer à la fin du Quaternaire: mesures, corrélations et applications prospectives" (projet PICG n°200) est venu apporter une dynamique incitative.

Les données demeurent encore inégales, mais aussi hétérogènes, et
des travaux importants et nouveaux comme ceux sur l'instabilité tectonique de certaines côtes, les facteurs de rupture d'équilibre de certains environnements margino-littoraux, l'hydrodynamique sédimentaire
des plages sableuses ou l'historique de la présence et de l'intervention de l'homme, sont le fruit de préoccupations initialement étrangères
à celle des lignes de rivage quaternaires. Ce bilan s'est efforcé de
prendre en compte de tels phénomènes dont les processus ont des conséquences directes ou indirectes sur la définition des fluctuations des
lignes de rivage anciennes ou plus récentes.

DEFINITION GEOSTRUCTURALE DU CONTINENT AFRICAIN

Un des caractères généraux de la structure du craton africain réside dans la pérennité, au cours des orogénies successives, des directions, des plis et cassures mises en place lors des mouvements précambriens. Lors de la formation des chaînes hercyniennes des Appalaches,
des Mauritanides et du Cap, il y a environ 250 millions d'années, le
bloc afro-sud-américain se trouva comprimé et sub-orthogonalement à
cette compression, des fissures sont intervenues selon des directions
souvent connues au Précambrien. La décompression postérieure aux dernières phases hercyniennes correspond à un rebond vers le Nord de l'A-
mérique du Nord (entraînant la Laurasie) et à l'ouverture de l'Océan
Atlantique central et à un rebond vers le Sud et le Sud-Est de l'Antarc-
tique et de l'Australie. Au niveau de l'Afrique centrale, la décompression permet, vers 195 millions d'années (fin du Trias), l'apparition des
granites du craton de l'Aïr. L'éclatement du Gondwana qui intervient
alors, conduit à l'ouverture de l'Atlantique Sud par sa partie distale
sud et à l'écartement des deux blocs africain et sud-américain avec
progression conjointe vers le Nord et le Nord-Est; la progression vers
vers le Nord-Est aidée par l'extension du fond océanique accentue l'ouverture de l'Atlantique Sud.

Il est admis que les failles du Rift oriental (King, 1970) furent
déterminées par d'autres plus anciennes, souvent précambriennes; elles
sont situées dans le quadrant Nord-Est tout comme le fossé d'effondre-
ment de la Bénoué interprêté (Grant, 1971) comme une fracture profonde

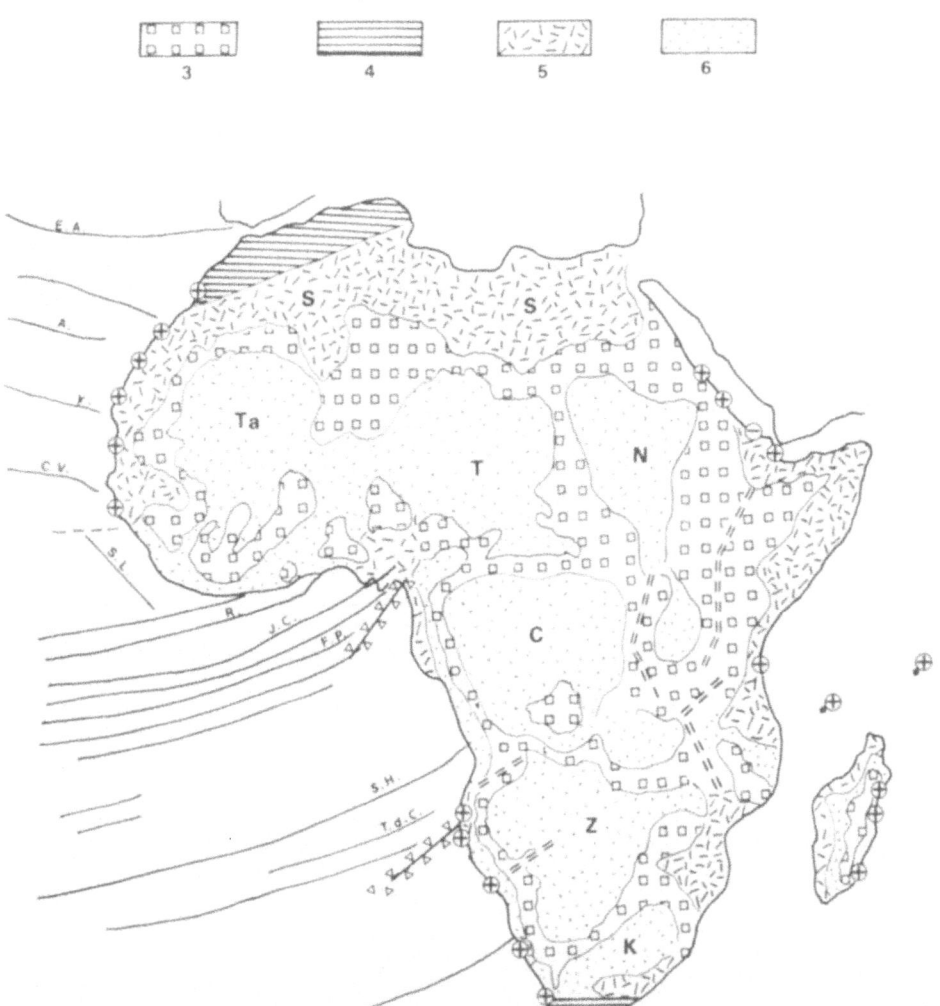

Fig. 1 - Cadre structural de l'Afrique. 1) - Rifts ou protorifts. 2) - Alignement d'antéclise. 3) - Zone de soulèvement du craton. 4) - Zones plissées de l'orogénèse alpine.
5) - Sédimentation péricratonique. 6) - Sédimentation intra-cratonique.
Fig. 1 - Structural framework of Africa. 1) - Rifts or protorifts. 2) - Anticline line.
3) Upliftcratonic area. 4) - Folded area of the Alpine orogenesis. 5) - Pericratonic sedimentation. 6) - Intra-cratonic sedimentation.

passagèrement en distension, puis comprimée plutôt que comme un rift
inactif; les Younger Granites du Nigeria (plateau de Jos et de Zinder)
témoignent de la phase de distension vers 160 millions d'années. Egalement, au droit de Walvis (zone de Kaako), l'alignement des basaltes cré-
tacés (125 millions d'années d'après Siedner et Miller, 1968) indique
une faille passagèrement de distension et orientée transversalement à la
côte. Le prolongement sous-marin de la Ride de Walvis est interprété
soit comme une fracture qui continue celle intra-cratonique (Le Pichon
et Hayes, 1971; Emery et al., 1975), soit comme le sillage de la plaque
Afrique au-dessus d'un hot spot (Burke et al., 1972).

La progression de la plaque Afrique vers l'Est, s'arrête aux alen-
tours de 80 millions d'années où le contact avec la plaque Eurasie se
traduit par un ancrage vraisemblable au niveau de l'Iran. Il en résulte
un nouveau mouvement vers le Nord-Ouest et finalement vers le Nord de
la plaque Afrique qui, conjugué à l'extension du fond océanique va pro-
voquer une ride de fond oblique de la lithosphère, grossièrement bissec-
trice des mouvements si les vecteurs vitesses ont le même module: le
volcanisme à basalte alcalin (trachyte et phonolite) prend naissance
sur cette ride médio-guinéenne vers 25 millions d'années et relaye ainsi les granites sub-volcaniques.

Cette ride médio-guinéenne que l'on peut qualifier d'antéclise joue
le même rôle de séparation des bassins sédimentaires que l'Anti-Atlas
paléozoïque au Nord du Bassin de Tarfaya -El Aaiun, que la dorsale Re-
guilhat entre celui-ci et le bassin Sénégalo-mauritanien ou que la ride
Lunda (prolongée sous la mer par celle de Walvis) entre le bassin de la
Cuanza et celui de Mossamedes. A l'échelle du continent, cette répartition antéclise-synéclise correspond à un agencement de structures beaucoup plus anciennes dont la dynamique serait pérenne. On observe, en
effet, au niveau de l'ensemble de l'Afrique une répartition suivant des
directions plus ou moins rayonnantes Nord-Nord-Ouest à l'Ouest jusqu'à
Nord-Nord-Est à l'Est d'alignements alternatifs de bassins et de bombements dont la longueur d'onde est de l'ordre d'environ 1250 km. Cette
répartition interfère avec celle de la succession grossièrement nord-
sud des bassins (Taoudeni, Tchad, Nil, Congo, Zambèze, Karoo) et des
reliefs de soulèvements antérieurs du craton. Nous aurons à considérer,
dans le cadre de ce bilan, cette bipolarité (S et NW) de l'origine des
antéclises et synéclises dont la tectonique intra-plaque est activée
pendant tout le Tertiaire par l'orogénèse alpine (Fig. 1).

Enfin, nous aurons à prendre en considération les terminaisons in-
tracratoniques sur la marge atlantique de l'Afrique de plusieurs failles
transformantes (failles de Saint-Paul, de la Romanche, du Chain, du
Charcot). Leur rôle peut être considéré comme déterminant en Guinée, au
Ghana, en Côte d'Ivoire, sites de séismes récents; la faille Saint-Paul
ordonne le bâti du socle côtier en Côte d'Ivoire et au Ghana, son prolongement sous le delta nigérien serait à l'origine du massif transver-
se de l'Adamaoua (Fail et al., 1970). Le rôle de ces failles semble davantage lié à des réactivations de la tectonique intraplaque qu'à un
réel prolongement du mécanisme transformant intra-océanique (Bellion et
al., 1984).

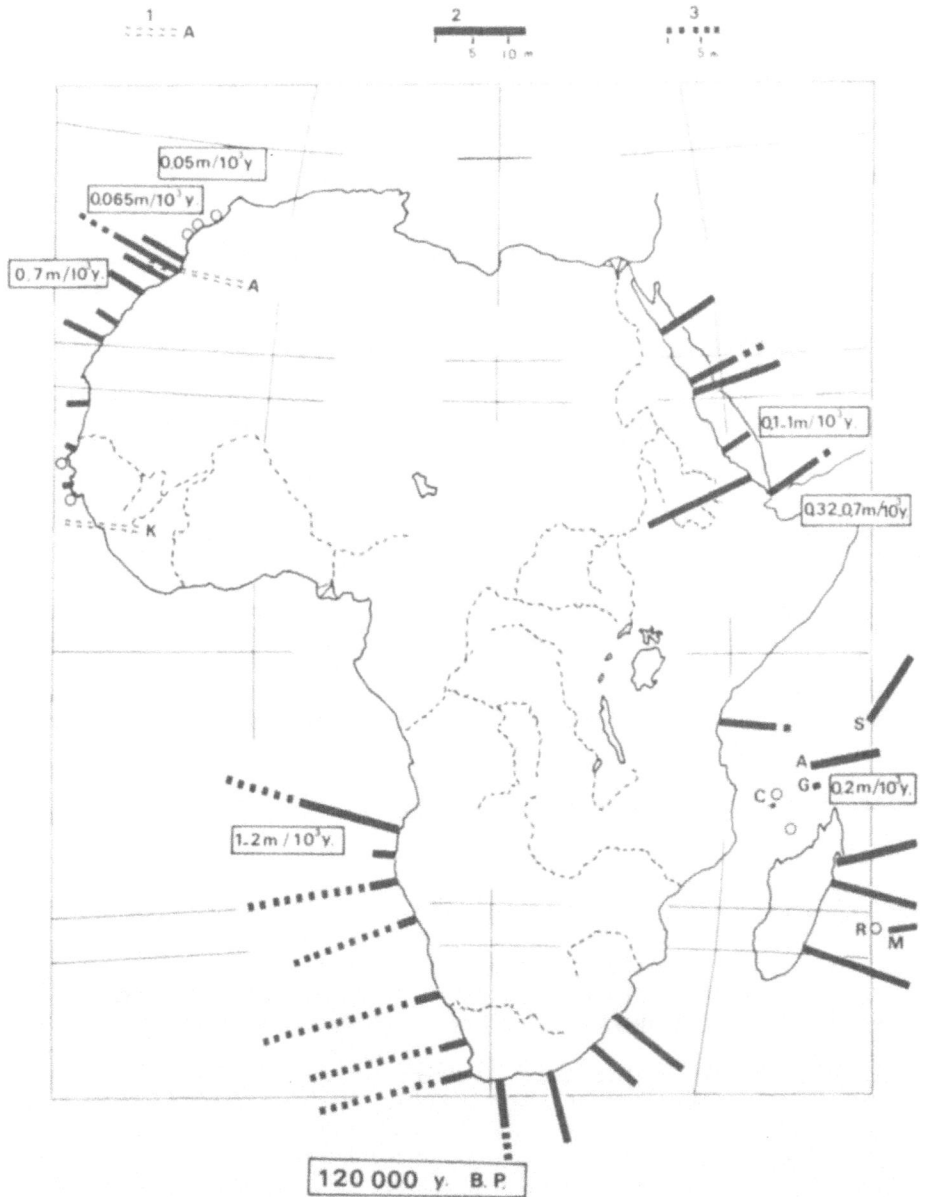

Fig. 2 - Carte des lignes de rivage de 120.000 ans. 1) - Flexure atlantique au Sud de l'Atlas. 2) - Altitude de niveau marin de 120.000 ans. 3) - Variante ou hypothèse d'altitude du niveau de 120.000 ans.
Fig. 2 - 120,000 y. B.P. shoreline map. 1) - Atlantic flexure, in the southern Atlas, 2) - 120,000 y. B.P. Sea-level. 3) - Other or hypothetical height.

LES LIGNES DE RIVAGE PLEISTOCENES DU CONTINENT AFRICAIN

La définition des lignes de rivage pléistocènes du continent africain est, aujourd'hui , assez inégale, dans la mesure où, bien des observations anciennes, qui sont les plus nombreuses, se sont avérées,
par la suite, plus ou moins douteuses. Sur les côtes du Golfe de Guinée, on a pu vérifier que certaines surfaces topographiques présumées
marines, résultent d'une ou plusieurs morphogénèses continentales con-
trôlées par les rejeux de vieux linéaments précambriens; d'autres fois
(Zaïre, Angola), les matériaux marins sont issus d'interventions anthropiques ou aussi, des remaniements colluviaux de matériaux antérieurs, parfois pliocènes (Angola, Afrique du Sud). Les secteurs côtiers où des mesures radiométriques répétées ont été effectuées (Maroc,
Sahara occidental et Mauritanie, Angola, Djibouti) ont mis en évidence
l'importance fréquente des recristallisations ou des échanges cationi-
ques au sein de carbonates issus souvent de tests de Mollusque; déjà,
des premières mesures préliminaires ont été l'objet de révisions com-
plètes (exemple de l'Harounien du Maroc). Ces difficultés méthodologi-
ques ne sont pas spécifiques au continent africain, mais en raison de
difficultés d'observations de diverses natures, elles prennent ici une
importance particulière au moment de ce bilan.
    Si l'on considère les cartes de lignes de rivage quaternaire du
globe qui furent publiées en 1967 et en 1981 par la commission des Lignes de rivage de l'INQUA, on est contraint, aujourd'hui, de présenter
un bilan beaucoup plus modeste où seuls les secteurs disposant de datations radiomètriques répétées seront retenus: parfois, et par extension,
nous prendrons en compte certaines terrasses marines caractéristiques
et à affleurement continu.

Le littoral occidental de l'Afrique

    Schèmatiquement, on peut distinguer des marges qui ont été le site
de mouvements épirogèniques positifs répétés dont le rythme et la cause
sont encore plus ou moins complètement analysés; il s'agit du Maroc, du
Sahara occidental et de la Mauritanie, au Nord, et de l'Angola, au Sud.
Elles s'opposent aux côtes atlantiques, entre le Sud du Sénégal et le
Nord de l'Angola où l'absence de témoins datés de niveaux marins pléistocènes surélevés indiquerait, très relativement, une plus grande sta-
bilité (Fig. 2).
    Le Maroc et, en particulier, le Maroc méridional, constitue le
secteur margino-littoral de l'Afrique où les témoins marins pléistocènes
sont les mieux connus. On suppose, au Nord du Cap Sim, une côte où les
transgressions se recouvrent et s'effacent les unes les autres et, au
Sud, une côte où sont individualisés plusieurs hauts niveaux océaniques
(Table I). Le mouvement épirogénique positif (0,5 à 0,7 m/10^3 ans)
coïncide avec les reliefs de l'Atlas dont l'érosion conduit à un relè-
vement isostatique qui atteint son maximum au niveau d'Agadir, c'est-à-
dire à l'emplacement de la flexure atlantique. On peut suivre l'affaissement de ces hauts niveaux, vers le Nord, entre le Cap Sim et Safi où
l'Ouljien I (125.000 ans environ) passe de (+8 à +12m) à (+5 à +6m) et
vers le Sud, où ce même Ouljien est à +4m. Ce rythme d'émersion est

encore plus lent à hauteur de la Meseta marocaine, les calculs à l'é-
chelle des trois derniers millions d'années indiquent 0,065 m/10^3 ans
dans le secteur de Casablanca et 0,053 m/10^3 ans dans celui de Rabat
(Stearns, 1978).

	ATLAS	TARFAYA	SAHARA OCCIDENTAL	MAURITANIE	SENEGAL
HOLOCENE	Mellahien + 1, + 2 m 6.000 - 1.000 y.B.P.		+ 1, + 2 m	Dakarien (+ 1m) 2.000 y. B.P. Taffolien (- 6m) 1.000 y. B.P. Nouakchottien 5.500 y. B.P.	Dakarien (+ 1m) 2.000 y. B.P. Taffalien (- 2m) 3.000 y. B.P. Nouakchottien (+ 2m) 5.500 y.B.P. Tchadien (- 20m) 7.000 y. B.P.
PLEISTOCENE SUPERIEUR	Ouljien III + 0,5 m 40.000 - 30.000 y. B.P. Ouljien II + 2 + 4 m 97.000 - 60.000 y. B.P. Ouljien I + 6, + 8 m 148.000-110.000 y. B.P.	Ouljien I + 4 m	Inchirien (?) Ouljien et Aïoujien + 2, + 5 m	Inchirien (?) Aïoujien (0 m) Tafaritien (?)	Aïoujien (0 m)
PLEISTOCENE MOYEN	Harounien-Rabatien,+20m 260.000 y. B.P. Anfatien, +20, + 30m 0.52-0.34 M.y Maarifien, +20, +50m 0.6-1 M.y.		? ? ?	BRUNHES ——0.85 M.y.—— MATUYAMA	
PLEISTOCENE INFERIEUR	Messaoudien + 83,117 m 1.1-1.8 M.y. Fouartien + 180 m 2.4-2.8 M.y. Moghrebien + 360 m 3-4, 2 M.Y.	Messaoudien + 30, + 40 m Moghrebien + 50 m	Moghrebien(?)+30 m		

Table 1 - Essai de corrélation des terrasses marines des côtes nord-ouest du continent
africain.
Table 1 - Attempt of correlation for transgression cycles in northwestern Africa.

Au site d'Agadir, la vitesse épirogènique est assez importance pour
permettre des corrèlations avec les cycles isotopiques de l'oxygène du
plancton marin, une succession très complète de niveaux marins peut y
être observée (Weisrock, 1980, Weisrock, 1981, Brebion et al., 1984 et
Weisrock et al., 1985). Les datations atteignent jusqu'à 260.000 ans,
l'âge des niveaux plus anciens sont estimés par corrélation (Table I).
Trois cycles marins sont reconnus dans le Pléistocène inférieur (Moghrébien, Fouartien et Messaoudien), neuf cycles transgressifs sont dis-
tingués dans le Pléistocène moyen et sont regroupés en trois étages
principaux (Maarifien, Anfatien et Harounien-Rabatien) et le Pléistocène supérieur se compose de la succession des Ouljiens I, II et III. Les
lignes de rivage les plus anciennes peuvent être moins élevées que celles plus jeunes, celà signifierait que depuis 0,85 millions d'années
(passage de Matuyama à Brunhes), les maxima transgressifs voisins du
zéro actuel ont coïncidé avec une accélération de la vitesse d'émersion,
notion donc différente de celle d'une émersion régulière présumée antè-
rieurement (Stearns, 1978). Les réchauffements des Interpluviaux correspondent aux principales phases transgressives, mais sont d'intensités
inégales, les eaux du Moghrébien et de l'Ouljien furent particulièrement

chaudes tandis que celles du Maarifien, de l'Anfatien et de l'Harounien furent relativement plus tempérées. Les corrélations paléontologiques avec les niveaux méditerranéens demeurent hypothétiques en raison du rapport élevé des faunes sénégalaises aux faunes lusitaniennes, le Fouartien est mis en équivalence avec le Plaisancien, le Messaoudien avec le Calabrien et l'Ouljien avec le Tyrrhénien.

Plus au Sud, la domination de la faune guinéenne (Brebion et Ortlieb, 1976) conduit à des comparaisons directes plus difficiles. A Tarfaya, le Moghrébien descend à +50m, le Messaoudien est réduit au seul gisement de Sidi bou Malet (+30 à +40m) et l'Ouljien, vers +4m, est encore hypothétique. Ce mouvement d'affaissement s'accentue au Sahara occidental, une dalle lumachellique (Moghrébien) observée à +30m disparaît vers 23°N et l'Aïoujien (équivalent de l'Ouljien I et II) est suivi entre +2 et 5m sur plus de 100 km de côte.

En Mauritanie et au Sénégal, les niveaux marins pléistocènes sont très faiblement étagés et jalonnent des rivages de golfes transgressifs (Elouard, 1976). Cette quasi-absence d'épirogènie rend très improbable la présence au-dessus du zéro océanique actuel de témoins datés vers 35-40000 ans B.P. que l'on observe, par contre, enfouis de -20 à -30m. Des travaux en cours, par ailleurs, tendent à conclure que les dépôts du Tafaritien ne sont que des accumulations du domaine continental. En conséquence, il est probable que la majorité des niveaux de beach-rocks très faiblement surélevés de ces côtes appartiennent à une ou plusieurs lignes de rivage du dernier interglaciaire (Eémien); la fréquence et l'importance des recristallisations des tests comme des ciments récemment constatées (Diouf et al., 1986) ne permettront qu'exceptionnellement les datations radiométriques. La stratigraphie du Quaternaire marin de ces régions souvent citée en référence appelle donc une révision importante.

Au Cameroun, les grands cordons sableux de Tiko à +20m viennent d'être assimilés à l'Eémien (Battistini et al., 1983), les auteurs attribuent à des "milieux de lagune ou de fonds de baie abrités" des dépôts d'argiles litées à passées sableuses. De telles accumulations sont considérées comme des dunes littorales pléistocènes sur des côtes voisines, et, en l'absence de mesures radiomètriques, ce problème demeure non résolu.

Sur la côte sud de l'Angola, les auteurs portugais (Soarès de Carvalho, 1961, en particulier) ont défini une terrasse de l'"Ouljien" vers +20m et du "Tyrrhénien" vers +50m. La terrasse de +20m est le résultat de la répétition de l'action de plusieurs hauts niveaux océaniques. L'érosion différentielle a pu préserver des sédiments du même âge, soit à +5m, soit à +15m sans qu'il soit nécessaire de faire intervenir des différences de tectonique locale; les mêmes observations peuvent être faites à propos de la haute terrasse de +50m. Dans le bassin de Lobito-Benguela, la terrasse de +20m supporte à sa base (+8 à +10m) des témoins marins de 36.000 ans B.P. (U/Th) et plus haut (+12 à +25m), différents dépôts du dernier interglaciaire que l'on peut corréler avec les Eémien II et III. Par contre, sur le littoral de Mossamedes, à la même altitude, on observe des dépôts marins beaucoup plus anciens de l'interglaciaire Riss-Mindel, qui indiqueraient une activité moins intense que dans le secteur de Lobito-Benguela. Cette faiblesse d'épirogénie s'accentuerait sur les côtes de Namibie où les niveaux de +2 à 4m sont attribués, bien que non datés, au dernier interglaciaire; dans ce secteur, les terrasses diamantifères du Pléistocène n'ont fait l'objet, ni d'étude fondamentale, ni de datations, mais la permanence remarquable d'un haut niveau entre +15 et +20m retient l'attention, il pourrait s'agir du véritable haut niveau éémien. Malgré le nombre toujours trop faible de datations, on peut estimer, à Lobito, une vitesse épirogènique des niveaux éémiens qui serait de l'ordre de 0,1 à 0,2m/10³ ans; elle est inférieure à 0,1m/10³ ans dans les dépôts plus anciens qui intègrent un plus grand nombre de phases de repos dans leur bilan (Giresse et al., 1984).

Le littoral oriental de l'Afrique (Table II).

L'Afrique du Sud montre un étagement remarquable des niveaux marins quaternaires: les hautes terrasses se trouvent à 400m (Port-Alfred), à 250m (Port-Elizabeth), à 120m (Natal), puis de manière moins nette à 110m, 95m, 82m et 73m; les moyennes terrasses sont générales vers 60m. Seules les basses terrasses sont datées par radiométrie et par les acides aminés: l'Eémien I (140.000 ans) est entre +6 et +10m, l'Eémien II (125.000 ans) entre +9 et +10m et l'Eémien III (85.000 ans) vers +12 à +14m. On constate donc que les dépôts de l'Eémien III (moins chaud) sont paradoxalement plus élevés que ceux de l'Eémien II (plus chaud); cette anomalie pourrait s'expliquer par une fonte précoce des glaces antarctiques par rapport à celles de l'hémisphère nord (Davies, 1981). Le dépôt de 125.000 ans à 8m est considéré comme de nature purement eustatique, un mouvement épirogènique ne semble donc pas intervenir sur cette côte.

Les rivages pléistocènes africains de l'Océan Indien sont encore mal explorés et rarement datés. Signalons au Kénya, la position moyenne entre 15 et 20m des niveaux du dernier interglaciaire dont l'altitude tend à s'élever en allant vers le Nord (Ase, 1982). A Djibouti, l'isochrone de 125.000 ans peut être suivi sur une grande distance, on constate les déformations inégales de panneaux étroits limités par des failles, le rythme des mouvements est très variable (0,1 à 1m/10³ ans) et discontinu dans le temps; le secteur du horst de Danakil paraît plus instable que celui plus au Sud (Faure et al., 1980 b). En Ethiopie, le paléorivage du dernier interglaciaire (âges au-delà de 70.000 ans) est formé de calcaires récifaux que l'on peut suivre sur 400 km de côte, le toit qui est +3, +5m à Massawa descend à -14, -18m à Assab (Faruque, 1986). Au Soudan (Dalongeville et Sanlaville, 1980), le récif fossile éémien vraisemblable ne dépasse pas +6 à +7m, le sommet étant souvent érodé, il est admis que le niveau océanique responsable a pu atteindre +10m, mais qu'il ne soit pas nécessaire de faire intervenir une néotectonique active. Enfin, sur la côte égyptienne de la Mer Rouge, un travail déjà ancien (Butzer, 1972) décrit dans la région de Mersa Alam des récifs coralliens à +6m qui sont datés à 80.000 ans B.P. et des coquilles de plages graveleuses datées à 118.000 ans B.P.

La côte orientale de Madagascar (Table II) montre une succession de terrasses de l'Eémien, ici appelé Karimbolien, qui sont datées à 150.000 ans B.P. entre +10 et +15m, à 120.000 ans B.P. entre +8 et +12m

AFRIQUE DU SUD R. Klazies	MADAGASCAR	MASCAREIGNES	SEYCHELLES	MOZAMBIQUE	DJIBOUTI	SOUDAN
+ 1 m 2540-4850 BP + 2 m	+ 1,5m 5500-6000 BP	0m 2000 BP	≃ 0m 3700-4300 BP	+2,5m-5-6000 BP		+0,6 m
+ 14 m Eem III + 12 m 85.000 BP	↑ + 3m 80-100.000 BP	32.000 BP ≃ 0m	+8M 120.000BP			
+ 9 m Eem II + 10 m 125.000 BP	Karimbolien					
+ 8 m Eem I 140.000 BP (Middle Stone Age)	120.000 BP ◄——— ——► 120.000 BP (Ivononien) ≃ 0 m				124.000 BP	+ 6,7 à +10 m
	↓ +15 m 150.000 BP					
+ 18 m (anté 140000 BP (Acheuléen)	Tatsimien ◄—— — —► 230.000 BP ≃ 0 m 350.000 BP ≃ 0 m 440.000 BP ≃ 0 m 470.000 BP ≃ 0 m					

Table 2 - Essai de corrélation des terrasses marines des côtes africaines de l'Océan Indien.
Table 2 - Correlation of transgression cycles in African coasts of the Indian Ocean.

et à 80-100.000 ans vers +3m. Cet étagement ne pose donc pas le problème de l'Afrique du Sud d'un Eémien récent qui serait plus élevé que les Eémiens plus anciens (et plus chauds) (Battistini, 1978). Par contre, les îles voisines des Comores ne présentent aucun niveau marin pléistocène surélevé. Dans les îles Glorieuses (Gaven et Vernier, 1979), les niveaux datés entre 100.000 et 108.000 ans affleurent entre +1 et +4m et conduisent à envisager, depuis 130.000 ans, un phénomène d'épirogénie positive de l'ordre de $0,2m/10^3$ ans. Plus au Nord, dans l'atoll d'Aldabra, seuls les coraux de 125.000 ans sont observés et traduisent une épirogénie plus importante (Thomson et Walton, 1972). Il en est de même dans l'archipel des Seychelles où le niveau de 125.000 ans est à +8m. Pour Gaven et Vernier(1979), on connaît, dans le canal de Mozambique des îles hautes, telles les Comores, où l'on note une activité volcanique et pas de récifs coralliens surélevés et des îles basses, telles les Glorieuses, où il n'y a plus d'activité volcanique mais où on observe, en revanche, des récifs en altitude; au niveau des îles hautes, le fond volcanique aurait une tendance à l'émersion en raison des émissions de matériaux éruptifs, les îles basses seraient des exemples où l'arrêt du volcanisme et le relachement des contraintes exercées entraînerait un mouvement de surrection.

Dans les Mascareignes, des niveaux témoignent de maxima marins vers 470.000, 440.000, 350.000, 230.000 et 120.000 ans §datations U-Th). Les

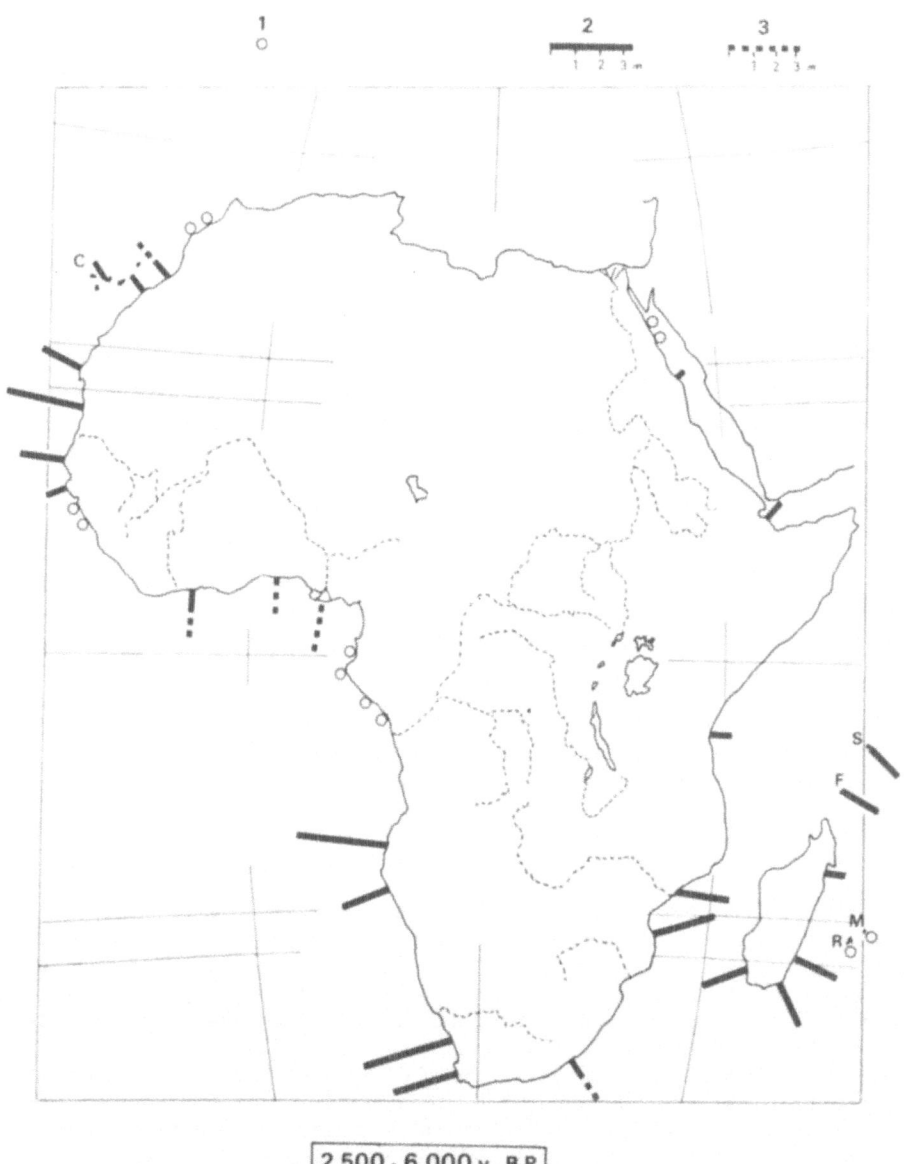

Fig. 3 - Cartes des lignes de rivages de 2500-6000 ans B.P. 1) - Absence de ligne de rivage émergée. 2) - Altitude du niveau marin de 2500-6000 ans B.P. 3) - Altitude hypothétique.
Fig. 3 - 2500-6000 y. B.P. shoreline map. 1) - No emerged shoreline. 2) - 2500-6000 y.
B.P. sea-level heights. 3) - Hypothetical height.

îles de la Réunion et Maurice sont relativement subsidentes, mais les accumulations coralliennes permettent de conclure à une remontée relative du niveau marin de 2,7m/1000 ans à Maurice qui est une île relativement stable et d'âge miocène et de 4m/1000 ans à la Réunion, plus instable et subsidente et d'âge seulement pléistocène.

LES LIGNES DE RIVAGE EMERGEES HOLOCENES DU CONTINENT AFRICAIN.

Les dernières étapes de la transgression holocène ont pu être observées sur la plupart des marges africaines, mais elles sont particulièrement reconnues à ce jour sur la façace atlantique. Les niveaux dont l'altitude est supérieure à celle du niveau actuel de l'Océan présentent une répartition qui dans ses grandes lignes, ressemble à celle des niveaux pléistocènes surélevés (Fig. 3). Ainsi, on observe des dépôts de plage holocène dans les secteurs allant d'Agadir jusqu'au Nord du Sénégal, de l'Angola à l'Afrique du Sud, sur les côtes de l'Océan Indien, sur la côte orientale de Madagascar et sur les côtes de certaines îles volcaniques non subsidentes (Seychelles, Farquhar). Les côtes du Golfe de Guinée sont généralement dépourvues de niveaux holocènes supérieurs au niveau océanique actuel, quelques secteurs plus instables comme ceux du Ghana, de la Côte d'Ivoire ou du Cameroun peuvent faire exception, mais la définition de leurs témoins demeure parfois encore discutable. Nous considérerons, plus en détail, les exemples du Sénégal, de la Côte d'Ivoire et du Congo.

Au Sénégal, à partir de 8400 ans B.P., un ralentissement de la transgression et un environnement humide sont favorables à l'installation des mangroves. La transgression atteint son maximum entre 6800 et 4200 ans B.P., c'est le Nouakchottien qui culmine entre +1 et +2m au Sénégal et à +3m en Mauritanie. Puis, de 6500 ans à nos jours, on va enregistrer une série de pulsations positives de +1 à 1,5m ou négatives de −1 à _2m dont la signification peut être très locale. Dans le delta interne du Sénégal, les principaux maxima marins sont datés vers 6700, 6200, 5600, 3400, 1850 et 1550 ans B.P. (Monteillet et al., 1981). L'oscillation négative la plus importante est intervenue vers 4000 ans B.P. où une régression de −2m (Taffolien) a contrôlé la fermeture du golfe du delta du Sénégal; dans ce site particulier, les lignes de rivage résultent de l'interfèrence des mouvements eustatiques avec les variations interannuelles, décennales et séculaires du débit du fleuve Sénégal, par exemple, l'extension dans le Ferlo, de faciès saumâtres vers 3300 ans B.P. et entre 1800 et 1500 ans B.P., correspond à deux périodes où l'écoulement du fleuve Sénégal et le niveau de la nappe phréatique ont été supérieurs à l'actuel (Fig. 4). En allant vers le Sud du Sénégal, la Gambie et la Casamance, les témoins marins holocènes deviennent inférieurs ou égaux au niveau actuel, (Faure et Hébrard, 1977). Cette différence latitudinale à l'échelle de 6000 ans (Fig. 5) implique une déformation relative apparente de l'ordre d'un mètre par 1000 ans dans une zone côtière relativement stable. Ce rythme de déformation est d'un ordre de magnitude une à deux fois supérieur à celui observé dans les accumulations cénozoïques et implique donc l'interférence des facteurs d'origine hydro-isostatique ou géoïdale à l'échelle réduite des phénomè-

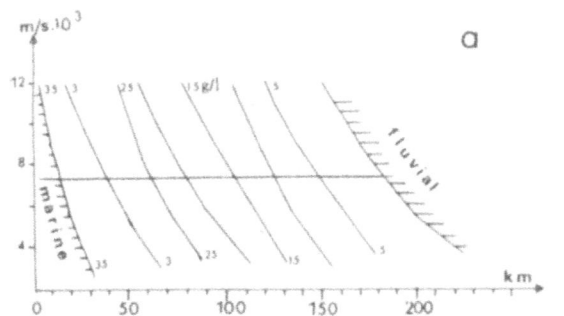 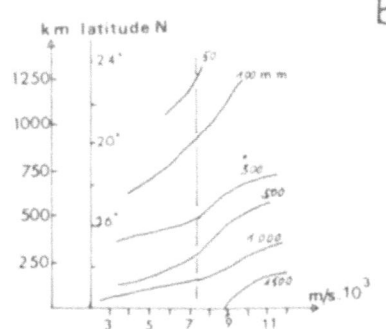

Fig. 4 - a) - Remontée de la langue salée (km) en fonction du débit (m³/s) du fleuve Sénégal; b) - Relation entre le déplacement en latitude des isohyètes annuels (mm) et le débit moyen du fleuve Sénégal (m³/s) (d'après Gac et al., 1981).
Fig. 4 - a) - oscillation of the salt water tongue position as a function of the river run-off; b) - Relation between the latitudinal displacement of annual isohyets (mm) and the run-off (m³/s) of the Senegal river (after Gac et al., 1981).

nes eustatiques observées (10^3 à 10^4 ans). Enfin, le bassin du fleuve Sénégal dont la plaine alluviale voisine du zéro océanique s'étend sur une distance de près de 150 km à la côte a été l'objet de mesures indirectes de la flexuration de la lithosphère (projet Rhéomarge). Les modèles calculés par Clark et Bloom (1979) indiquent une flèche verticale de l'ordre de 5 à 10 m en quelques millénaires sur quelques centaines de kilomètres mesurés perpendiculairement au littoral. Ici, la déformation n'est que de 2m (environ) depuis 7000 ans et d'un mètre depuis 6000 ans (Fig. 6). En conséquence, la rigidité du manteau ou de la lithosphère sous cette partie de la marge de l'Afrique de l'Ouest est plus importante que celle admise dans les modèles précédemment étudiés de l'Australie et du Brésil (Faure et al., 1980 a).

En Côte d'Ivoire, la relative abondance des pollens et des spores issus, soit de la mangrove, soit de la forêt marécageuse, permet de mettre en évidence quatre phases transgressives pendant l'histoire des cinq derniers millénaires de l'ancien delta de l'Agnéby qui se jette dans la lagune Ebrié. (Fredoux, 1983). L'auteur oppose les marqueurs de la forêt marécageuse à ceux de la mangrove dont les fluctuations respectives sont à peu près inverses. Il est à remarquer que les Fougères Filicinées s'accommodent bien de l'expansion de la mangrove et ne semblent pas gênées par le milieu saumâtre. Seul le niveau de base a été daté (-5m à

134

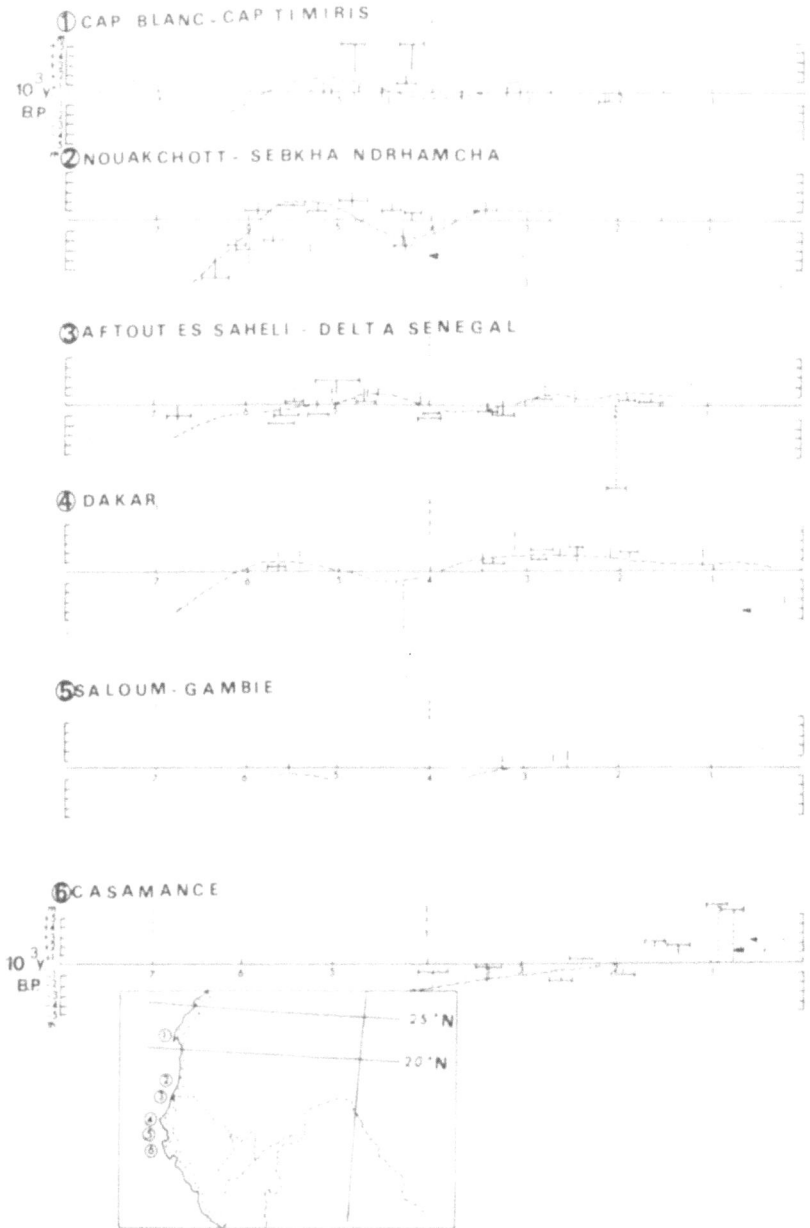

Fig. 5 - Lignes de rivage holocène de l'Afrique de l'Ouest en fonction de la latitude; le tracé est une ligne de rivage possible basée sur les observations de terrain et les analyses écologiques et sédimentologiques (d'après Faure et Hébrard, 1977).
Fig. 5 - Altitudinal sea-level in various latitudinal areas of West Africa; the line drawn is a suggested possible mean sea-level position based on field observation and ecological and sedimentological deduction (after Faure et Hébrard, 1977).

4990 ans), il permet, par extrapolation, de proposer les âges de 3750, 2750 et 1750 ans pour les pulsations positives successives du niveau marin, la plus nette de ces pulsations étant la seconde. Si on se réfère aux études sédimentologiques et morphologiques antèrieures (Tastet, 1979), le premier développement de la mangrove correspond au premier haut niveau marin après 6000 ans B.P. (où se construisent les cordons blancs) et les trois pulsations postèrieures se situent dans la phase de transgression entre 4000 et 1000 ans B.P. (où s'édifient les cordons roux de 2° génération).

Au Congo, vers 8000 ans B.P., d'importants restes de tourbes de mangrove découverts vers -20m par sondages dans le sous-sol de Pointe-Noire témoignent du maximum de cette végétation littorale. Les chenaux sont encaissés au milieu d'un paysage arbustif de mangrove. Puis, le paysage devient plus instable, les chenaux sont divagants: les accumulations tourbeuses lenticulaires passent latèralement aux sables fins des chenaux ou, dès 5000 ans B.P., aux sables marins coquilliers (Giresse, 1981). A partir de 5000 ans B.P., l'océan a pratiquement atteint son niveau actuel, les oscillations postèrieures sont de petite amplitude et résultent de réajustements hydroisostatique dont la chronologie ne peut être qu'esquissée: faible régression (moins d'un mètre) vers 4000 à 3000 ans B.P. avec des actions de déflation éolienne, deuxième maximum marin entre 2000 et 1500 ans B.P. qui a pu dépasser le zéro actuel (+0,5m?) et correspondre au maximum d'extension des lacs et marécages côtiers et léger mouvement négatif qui ramène le niveau de l'océan au zéro actuel et, vraisemblablement restreint la zone intertidale habitée par la mangrove.

REMARQUES SUR LA GEODYNAMIQUE DU CONTINENT.

Les marges ou paléomarges sont les lieux où s'inscrivent le plus souvent les marques des changements relatifs du niveau de la mer. Ces marques ou isochrones témoignent de variations plus ou moins importantes dans l'espace des niveaux relatifs de l'océan qui impliquent notamment des déformations tectoniques de nature locale ou globale. Nous avons dans le cadre de ce continent à considérer des marges de type "atlantique" ou de type "indien" que l'on peut qualifier de "passives" ou de "stables".

On constate, au niveau de l'ensemble de l'Afrique, des alignements grossièrement Nord-Sud des bassins et des bombements alternatifs, cette répartition suit des directions plus ou moins rayonnantes Nord-Nord-Ouest à l'Ouest et Nord-Nord-Est à l'Est. Transversalement à ces alignements, suivant des courbes concentriques des successions alternatives d'antéclise et de synéclise (Fig. 1). La ride médio-guinéenne constitue une antéclise étroite qui est accompagnée d'une synéclise au large du Congo-Cabinda caractérisée en sismique par une réflexion du socle prémésozoïque qui est nettement plus basse qu'au niveau du Nord Gabon et de l'Angola. Plus au Sud, nous retrouvons une antéclise majeure avec la ride de Walvis. Ces synéclises semblent jouer actuellement encore car on constate, au Sud du fleuve Congo comme au Nord de la Casamance, des épirogènies positives qui affectent tant les témoins marins pléistocènes

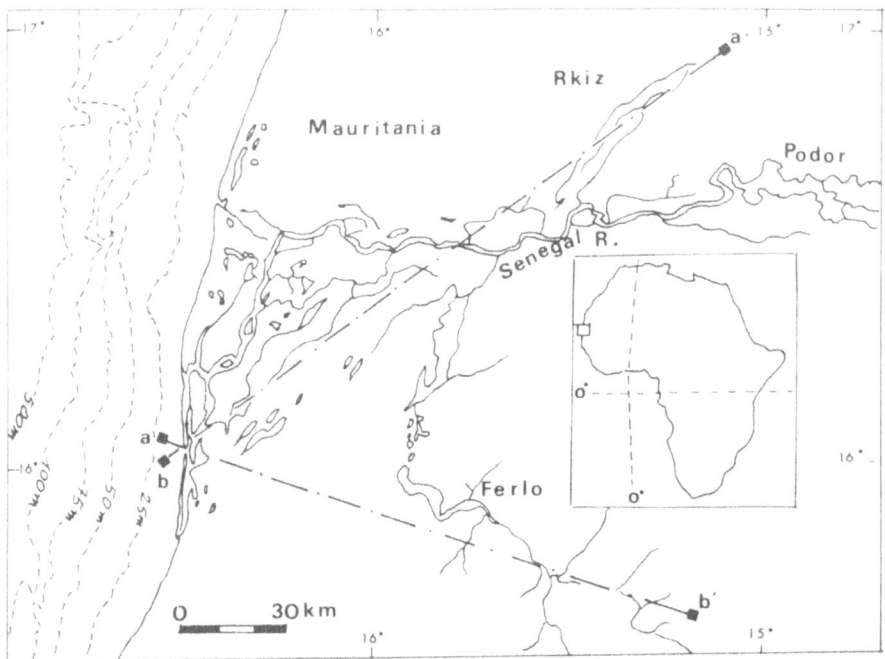

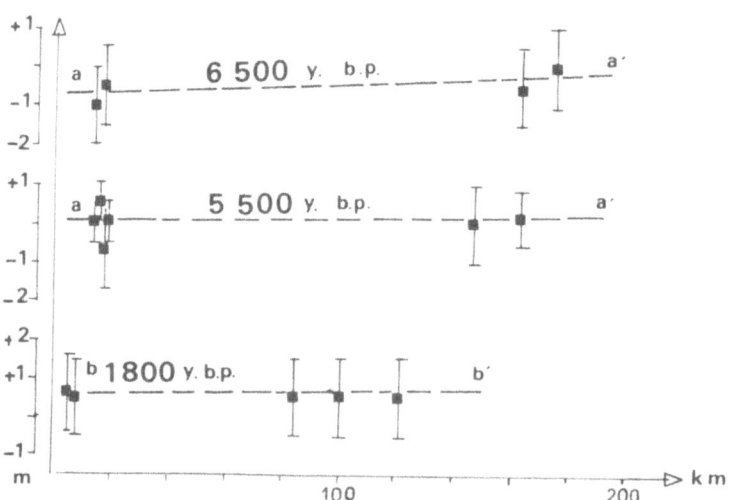

Fig. 6 - Reconstitution de la surface possible du geoïde à 6500, 5500 et 1800 ans B.P. le long des traverses AA' et BB' à travers le delta du Sénégal et le Ferlo (d'après Faure et al., 1980 a).

Fig. 6 - Portions of the geoïd surface dated 6500, 5500 and 1800 y. B.P. along AA' and BB' through the Senegal Delta (after Faure et al., 1980 a).

qu'holocènes. Elles semblent préserver une disposition bipolaire et être organisée autour de centres définis au Nord et au Nord-Ouest du continent, d'une part, et au Sud de celui-ci, d'autre part. Leurs effets ne semblent pas enregistrés dans la partie centrale de l'Afrique.

L'échelle moyenne des mouvements peut atteindre jusqu'à un mètre sur 1000 ans le long de 1000 km de côtes et également d'un mètre sur 1000 ans en s'éloignant de 100 km vers l'intérieur des terres. Ces valeurs sont une ou deux fois plus élevées que celles de la moyenne du Cénozoïque et expriment ainsi des phénomènes d'origine géoïdale ou hydroisostatique amplifiés depuis le Quaternaire par les mouvements eustatiques et les changements dans la répartition des masses d'eau et de glace qu'ils impliquent. Mais nous avons vu, dans le cas de la marge de l'Afrique de l'Ouest, que la rigidité de la lithosphère ou de l'asthénosphère semble indiquée par une déformation visco-élastique moins forte que celle mesurée en d'autres points du globe.

La marge de la Mer Rouge du continent est l'objet de nombreuses déformations qui affectent des panneaux étroits peu basculés, limités par des failles. Des zones épirogéniques positives comme le Rift de Danakil et le Golfe de Tadjoura sont voisines de zones plus stabilisées comme celle du Détroit de Bab el Mandeb entre le horst de Danakil et l'Arabie ou effondrés comme le secteur d'Assab; plus au Nord, le littoral égyptien semble plus stable.

Les mesures de taux de soulèvement depuis 125000 ans peuvent être souvent comparées à celles relatives à l'Holocène, elles soulignent le caractère discontinu des mouvements verticaux à plusieurs échelles de temps. Egalement, les mesures géodésiques et fournies par des nivellements répétés mettent en évidence le rôle épisodique des crises séismiques ou séismo-volcaniques fréquentes à l'Est de l'Afrique: l'éruption volcanique fissurale de 1978, a fait apparaître en Afar, un nouveau volcan, l'Ardoukoba: une distension brusque s'est traduite par un écartement de 1,60m dans l'axe du rift et par un soulèvement de 40cm environ. Mais aussi, en Afrique de l'Ouest, l'exemple du séisme récent de Koumbia (Guinée) qui a provoqué une surface de rupture de 15 km de long, témoigne d'une sismicité active, même si la résultante épirogénique n'est pas apparente; la même remarque peut être faite pour le littoral voisin du Mont Cameroun.

LES LIGNES DE RIVAGES SOUS-MARINES ET LEUR INTERET ECONOMIQUE.

Les plateformes du continent africain dont les sédiments superficiels ont fait l'objet d'une reconnaissance détaillée sont encore assez peu nombreuses, celles qui sont définies par une cartographie (Sénégal-Gambie, Côte d'Ivoire, Gabon-Congo-Cabinda-Zaïre) sont encore plus rares. Ainsi, on ne sait rien encore sur les plateaux du Libéria, du Ghana, du Bénin, de l'Angola, de la Tanzanie, du Kénya et de la Somalie.

L'observation à l'affleurement des lignes de rivage sous-marine est fonction de l'importance de la couverture sédimentaire, principalement vaseuse, qui est venue ensevelir (surtout après 6000 ans B.P.) les témoins des cordons sableux ou graveleux. Cette couverture récente est, de manière générale, particulièrement développée dans la zone intertro-

picale où, à l'échelle de la marge atlantique de l'Afrique, sont expor-
tés 76% du matériel terrigène du continent (Unesco, 1986). Egalement,
des zones d'apports sédimentaires très importants sont situés à la limite des zones arides et modérément humides où la végétation se raréfie,
l'exemple le plus significatif est celui du débit solide du fleuve Orange qui, bien qu'en zone aride, transporte tous les ans 150 millions de
tonnes à l'océan, soit davantage que le Congo et le Niger réunis. Au
nord, les alluvions de la Casamance et surtout du Rio Cacheu jouent ou
ont joué lors de l'Holocène des rôles comparables. Par contre, au large
des zones arides, l'apparition terrigène (surtout météoritique) est 15
fois moins important. Ces règles générales montrent de fréquentes ex-
ceptions dans la mesure où dans la zone intertropicale, il est fréquent
d'observer un piègeage alluvial dans les arrières-deltas des plaines
maritimes (cas de l'Ogooué) ou encore dans les biefs alluviaux des cours
moyen et inférieur (cas du Congo) ou encore le transport particulaire
directement vers les grands fonds océaniques: un dixième seulement de la
charge solide du Congo se dépose aujourd'hui sur le plateau continental
(Giresse, 1985). Ainsi, d'assez vastes surfaces des plateformes intertropicales montrent, aujourd'hui, à l'affleurement des témoins reliques
des lignes de rivage de la dernière transgression, c'est le cas, par
exemple,de la Guinée, de la Côte d'Ivoire, du Gabon et du Nord du Congo
(Fig. 7).
    Les bordures externes des plateformes de l'Afrique sont générale-
ment, couvertes, comme la plupart des plateformes des autres continents,
par des dépôts reliques qui correspondent au maximum de la dernière régression. Sur la marge atlantique de l'Afrique, la permanence du bas-
niveau marin entre 22.000 et 16.000 ans B.P. a eu pour conséquence une
action prolongée de la houle sur des fonds voisins de la rupture de pente qui a exercé un vannage des particules fines: ces dépôts sont essen-
tiellement des sables et dans le cas des marges intertropicales, des sables glauconieux auxquels une fraction vaseuse plus ou moins importante
peut-être associée (plateaux de Côte d'Ivoire, du Nigéria, du Gabon et
du Congo).
    Un autre caractère très général des dépôts des plateformes atlanti-
ques de l'Afrique réside dans l'accumulation d'un sable assez grossier
de bioclastes calcaires qui est observé entre -80 et -110m (voire -120m).
Cette "Amphistegina fauna" (Lagaaij, 1973), très riche en *Amphistegina
gibbosa,* se compose aussi de Mollusques et d'organismes littoraux d'eaux
claires: autres Foraminifères, Madréporaires, Algues calcaires et Bryo-
zoaires. Ces dépôts semblent correspondre à un unique prisme sédimentaire côtier largement remanié par les oscillations des houles de tempête
lors de l'étape de repos relatif de la transgression observée entre
13.000 et 11.000 ans B.P. L'origine de ce repos paraît eustatique à Morner (1976) qui propose une amplitude de 9 à 12 m pour une oscillation
négative entre 13 et 14.000 ans B.P. qui intervient après une première
phase de transgression rapide. Cette oscillation se traduit par une brè-
ve, mais importante extension des glaciers qui est associée à la fluctuation géomagnètique de Gothenburg (Fairbridge, 1977). Mais ce refroidis-
sement global n'a pas eu d'incidence sur la température des masses
d'eaux guinéennes dont la transgression sur le plateau va favoriser l'extension des moussons sur le continent. La majorité des organismes de la

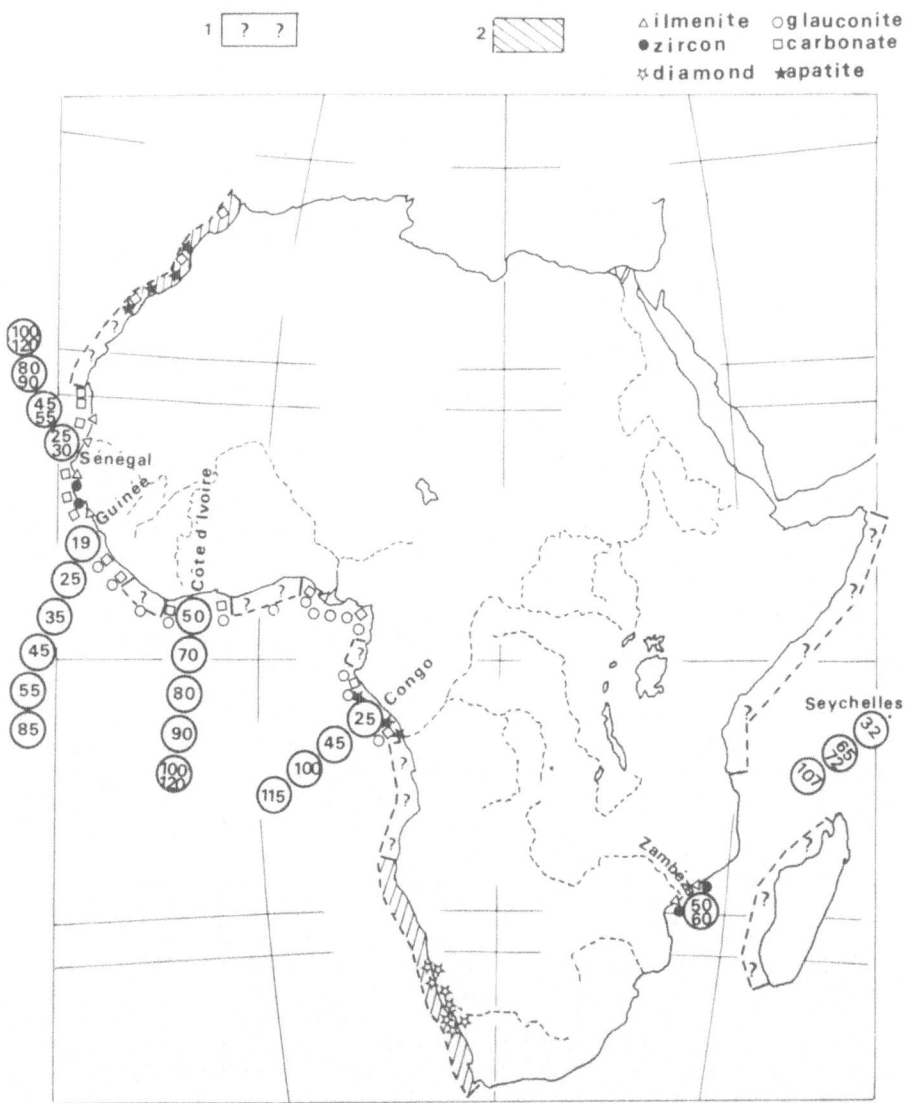

Fig. 7 - Témoins des lignes de rivage sous-marines du plateau et concentrations minéra-les; 1) - Pas de témoins de ligne de rivage. 2) - Plateau inexploré; la profondeur en mètres sous le zéro actuel est indiqué dans les cercles.
Fig. 7 - Holocene shorelines on the shelf and mineral accumulations. 1) - Shelves wi-thout shoreline indications. 2) - Unexplored shelf; numbers in the circles mean meters below sea-level.

faune et de la flore qui composent cette thanatocoenose vivait dans des
fonds dont la profondeur n'a pas excédé 30 à 50 m; la présence d'Algues
calcaires et d'Amphistégines qui vivent en symbiose avec les Algues
zooxanthelles et les coraux, indique la transparence d'une eau littorale où les fleuves n'apportaient encore que peu de matières particulai-
res à l'océan (Giresse et al., 1981). Ce dépôt a été daté plusieurs fois,
notamment au large de la Côte d'Ivoire (Martin, 1973) et du Congo (Giresse et al., 1984) entre 10.200 et 12.620 ans B.P. Au Congo, il constitue une couverture de 0,5 à 2 m d'épaisseur qui est répartie sur une
largeur de 20 à 25 km. Cette couverture de sable carbonaté meuble, éven-
tuelle ressource utile aux amendements des cultures sur sols tropicaux,
pauvres en bases échangeables, constitue un phénomène général qui est
vérifié sur la plupart des zones externes de la plateforme atlantique
de l'Afrique intertropicale (Barusseau et al., 1987).

De même, les bordures externes des plateaux africains (Sénégal, Guinée, Côte
d'Ivoire, Gabon, Congo, Seychelles) présentent une terrasse d'érosion
marquée entre -100 et -120m qui semble résulter plutôt de la répétition
des actions du déferlement des houles de bas-niveaux océaniques successifs que de celle, en particulier, de la dernière régression. Il semble
que les profondeurs les plus fréquentes de surface d'abrasion des plateaux de l'Afrique de l'Ouest, en général, soient, localisées près des
isobathes de 80 m (période 14500-20500 ans B.P.), de 50 m (période
11500-9500 ans B.P.),de 35m (période 9800-8500 ans B.P.), mais ces surfaces qui ne sont pas systématiquement présentes, ne sauraient avoir
une signification eustatique.

De même, les accumulations littorales abandonnées par la dernière
transgression sont assez diverses (Fig. 7). Au Sénégal, les ralentisse-
ments de la transgression ont conduit aux isobathes 90-80m, 55-45m et
25-30m, à des alignements de modes granulométriques de sables fins (Barusseau, 1984). Ces stationnements sont, par ailleurs, reconnus grâce
aux développements successifs de lagunes à ostracodes (Peypouquet, 1977);
ils sont datés autour de 11000 ans et 9000 ans et permettraient la définition de littoraux entre -50 et -40m et entre -30 et -20m. Au large
de la Guinée (Mc Master et al., 1971), plusieurs alignements de reliefs,
rides et dépressions correspondent à des vestiges de cordons littoraux,
d'îles barrières, de complexes lagunaires et de falaises marine, des
lignes de rivage sont supposées à -90, -80, -55, -45, -35, -25 et, éven-
tuellement entre -18 et -20m (Fig. 8); par ailleurs, sur la bordure ex-
terne des dunes littorales de régression ont été lithifiées à l'approche
de l'océan, puis mises en relief par érosion différentielle lors de la
transgression qui a suivi, ces barres gréseuses de plusieurs mètres sont
aussi connues au large du Sénégal. Au large aussi de la Côte d'Ivoire,
des alignements de dunes lithifiées ont pu être suivis sur plusieurs di-
zaines de kilomètres vers -90, -80, -70 et -50m. Au large du Congo et
du Gabon, les vestiges de cordons ou de dunes lithifiées sont très rares,
la distribution statistique des cailloutis ou des graviers soulignent
les profondeurs de stationnements répétés des lignes de rivage entre
-115 et -100m et entre -50 et -20m (Fig. 9); ces dernières auront une
incidence à propos de la définition de placers phosphatés. En résumé, il
paraît illusoire, pour l'instant, d'associer trop étroitement une profondeur donnée et une ancienne ligne de rivage, la fossilisation des

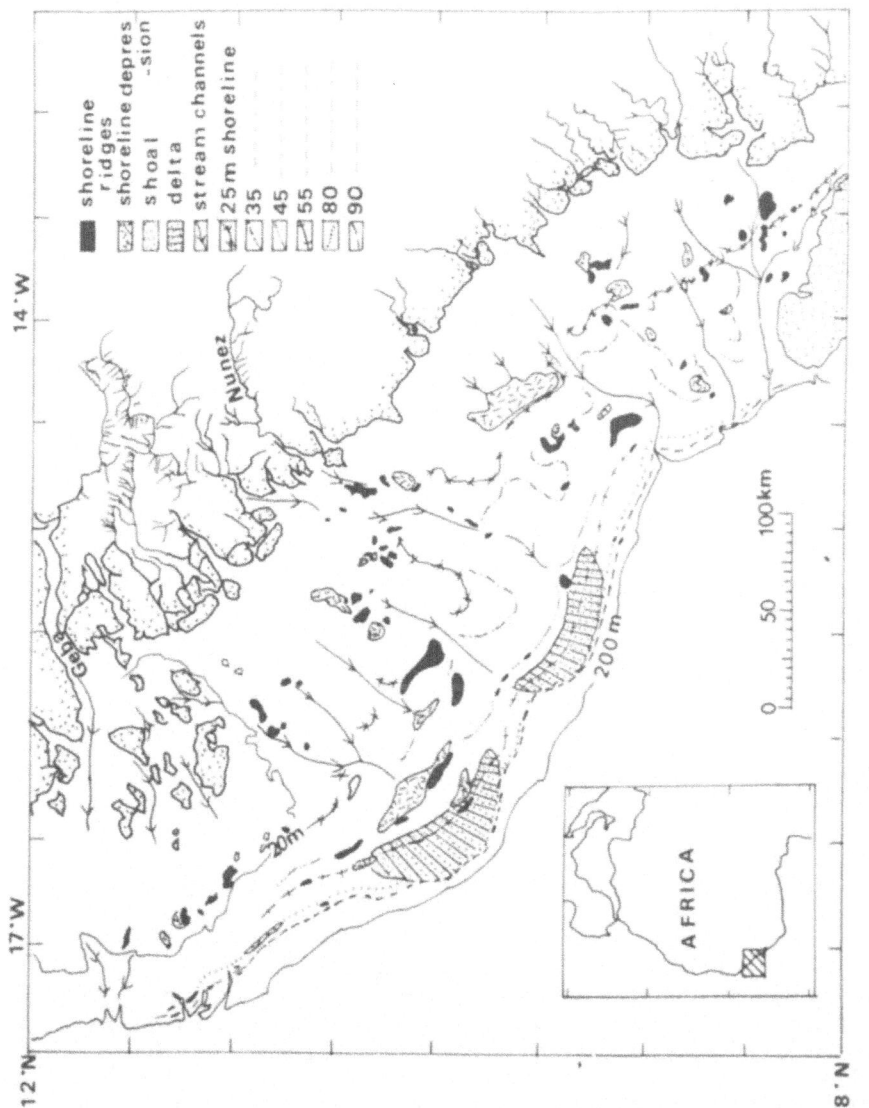

Fig. 8 - Les lignes de rivage sous-marines du plateau guinéen (d'après Mc Master et al., 1971).
Fig. 8 - Submerged shorelines of the Guinean shelf (after Mc Master et al., 1971).

142

témoins de celle-ci est alléatoire et son altitude a pu fluctuer en fonction de mouvements tectoniques postérieurs en dépôt.

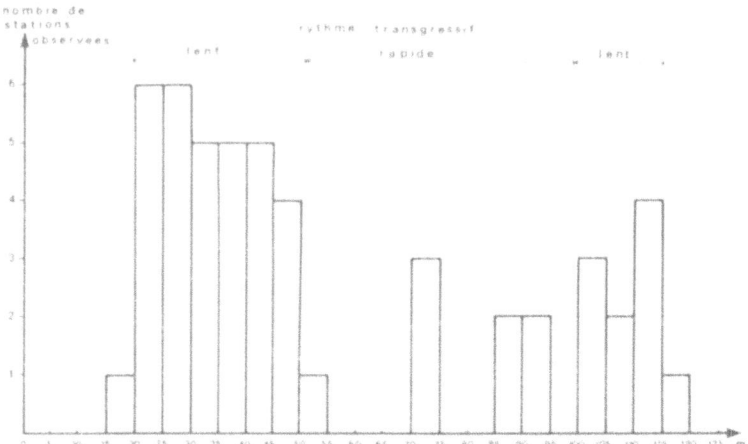

Fig. 9 - Distribution statistique des vestiges de lignes de rivage holocènes du plateau congolo-gabonais en fonction de la bathymétrie (d'après Giresse, 1981).
Fig. 9 - Statistical distribution of Holocene shoreline marks on the Congolo-Gabonese shelf related to the bathymetry (after Giresse, 1981).

Comme le niveau océanique actuel a été atteint, il y a 6000 ans et que le dernier haut niveau comparable remonte à 100000 ans environ, il y a eu une longue période où l'émersion a permis une importante accumulation de sables fluviatiles sur la plateforme. En zone intertropicale, les minéraux fragiles et instables comme l'amphibole, le pyroxène, le grenat, l'épidote tendent à disparaître alors que des minéraux plus résistants, comme le quartz, mais aussi plus utiles comme le zircon, le rutile, l'ilménite, la monazite, la magnétite, la brookite, l'anatase (et le diamant) se trouvent concentrés. Outre les concentrations liées aux placers de rivière où les particules en suspension ou en saltation se trouvent séparées de celles près du fond, les lignes de rivage sont le site d'un triage efficace: le déferlement amène un large spectre granulométrique de grains à la côte, un premier dépôt a lieu en fonction de la vitesse de sédimentation des particules; puis lors du mouvement de retrait, les grains légers et assez grands sont entraînés plus loin que les grains plus petits et plus denses, un film de minéraux lourds triés est constitué et la répétition de cette mécanique peut aboutir à des lits épais (Veenstra et Winkelmolen, 1976). Un tel processus a pu intervenir au large du Sénégal dans la formation des lits d'ilménites dont le premier agent de concentration a été la dérive littorale qui est venue remanier le sable éolien des ergs. Au Nord du Cap-Vert, un large dépôt d'ilménites entre -45 et -10m avec des concentrations entre -25 et -30m est lié aux lignes de rivage entre 8500 et 7000 ans B.P. Alors qu'au large de la Mauritanie (près du Cap Timiris) et de la petite côte du Sénégal (près de Mbodiène et de Joal), des placers de plage se sont constitués vers 5500 ans B.P. à l'arrivée de la transgression du Nouak-

chottien atteignant jusqu'à 8% d'ilmènites à Nianing (BRGM 1973, Pinson-
Mouillot, 1980; Barusseau et Giresse, 1987).

Au large de Djèno (Congo), les hauts fonds tertiaires riches en co-
prolithes de phosphate affleurent entre -35 et -50m et ont été atteints
et rapidement immergés par la transgression vers 9000 ans B.P. (Giresse et al., 1984). A partir de 7000 ans B.P., le rythme de la transgression a ralenti, et les convergences de la houle ont conduit à des remaniements et des vannages importants sous une tranche d'eau de 15 à 30m;
les concentrations de graviers de phosphate expriment une hydrodynamique
prépondérante de la dérive littorale et de la réfraction de la houle sur
les fonds immergés; les transits parallèles derrière et à l'abri du relief rocheux excluent un transfert notable des particules phosphatées
depuis le haut fond en direction de la côte. Vers 5000 ans B.P., l'océan
a atteint son niveau actuel et le vannage de la houle se déplace vers
la côte; dès lors, dans des eaux de basses énergies, débute la couverture par les vases alluviales du Congo.

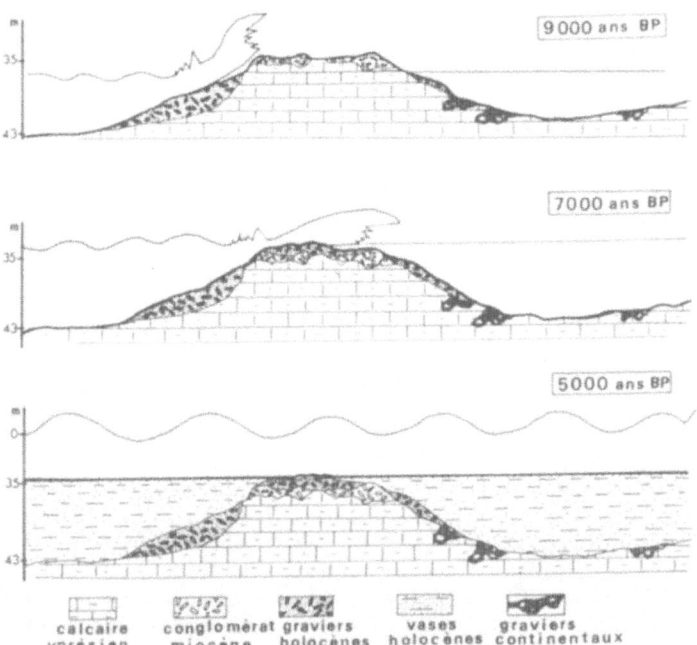

Fig. 10 - Etapes de la concentration mécanique des phosphates sur le haut fond de Djéno
(Congo): 9000 ans, le vannage est localisé sur la pente exposée au large, 7000 ans, le
remaniement est intense lors du ralentissement de la transgression, 5000 ans, les graviers phosphatés sont ensevelis en partie par les vases alluviales (d'après Barusseau et
Giresse, 1987).
Fig. 10 - Steps for mechanical concentration of phosphate on the bank 9000 y. B.P., the
winnowing action of the waves is located seaward of the bank; 7000 y. B.P., intensive reworking and winnowing action during the time when transgression was slowing down; 5000 y.
B.P., phosphorite gravels become thinly covered by barren muds (after Barusseau et Giresse, 1987).

Nous avons peu d'information sur les concentrations de diamants des plages actuelles et pléistocènes (surélevées et sous-marines) de la côte atlantique de l'Angola et de la Namibie (Namaqualand). Les diamants ont été amenés par le fleuve Orange au Sud et par plusieurs rivières côtières, au Nord (Bardet, 1974). Les diamants sont plus abondants en haut et bas de plage, mais aussi au fonds de marmites ou de ravinements du bed-rock qui se prolongent sous la mer; une plate-forme sous-marine à -20m est suivie sur 550 km de long et attribuée à l'Holocène (9000 ans?). Près de l'embouchure de l'Orange, deux terrasses pléistocène surélevées sont particulièrement exploitées: entre +2 et 10m et surtout entre +12 et +25m; en Namibie les gisements concernent surtout la terrasse au-dessus de +8m. Des rejeux tectoniques ont pu jouer et expliquer l'absence de terrasses diamantifères sous-marine au large de la Namibie et du Sud du fleuve Orange.

Sur le plateau du Mozambique, les sables deltaïques du Zambèze se sont largement étalés pendant la régression et ont permis la définition d'un vaste placer d'ilménites et accessoirement à zircons et rutiles (Kudrass, 1987). Les plus fortes teneurs sont définies dans des lentilles sableuses holocènes entre les isobathes 50 et 60m. La transgression holocène a pu localement, le long du cordon du front deltaïque, aboutir à des concentrations de placer, mais le grand volume de sable n'a pu être entièrement entraîné vers la côte pendant la suite de la transgression et le dépôt a été plutôt élargi et les concentrations dispersées pendant un trajet qui fut long et rapide. Comme pour le plateau du Sénégal, on peut considérer que les sables minéralisés peuvent être assez largement étalés , mais sans que soit effacés les domaines paléogéographiques de leur dépôt initial. Le modèle du Zambèze implique un plateau large, des apports modérés à forts en particules terrigènes et une côte à haute énergie. Pour Kudrass (1987), il s'oppose à celui de l'Ouest de l'Australie où sur un plateau étroit doté d'apports terrigènes faibles à modérés et d'une côte à haute énergie, le cordon littoral pourra migrer à travers le plateau pendant la transgression avec une teneur en minéraux denses et utiles continuellement concentrée. Le modèle du Zambèze paraît plus représentatif des plateaux africains, mais peut présenter plusieurs variantes liées à la distance à la côte de l'accumulation minérale à l'origine de la concentration future des ilménites du Sénégal (proche de la côte) ou des bioclastes calcaires (éloignées de la côte) ou au grain moyen des minéraux utiles (cas des graviers de phosphates du Congo) qui n'ont pas été entraînés par la transgression) (Fig. 10).

La transgression holocène ne concentre et n'accumule (rôle de vannage) que lors de ses phases de stabilisation ou de ralentissement, elle ne charrie (rôle de bulldozer) que sur des distances assez courtes et si son mouvement positif n'est pas trop accéléré.

LES LIGNES DE RIVAGE ET LES SUCCESSIONS DES OCCUPATIONS HUMAINES

La proximité des lignes des rivages a déterminé souvent la succession des occupations humaines.

A ce jour, aucune étude ne permet d'établir les relations des premières populations humaines et pré-humaines avec le rivage de l'Océan

Indien. Les civilisations côtières les plus anciennement connues sur
les côtes de l'Afrique seraient celles de l'Angola: australo-anthropiens
à la limite du Pléistocène inférieur et moyen (Ramos, 1981), industrie
d'un Oldowayen peu évolué décrit dans la Lunda (Ramos, 1982), puis développements successifs au Pléistocène moyen d'un complexe acheuléen
lié à des populations d'Archanthropiens (Pithécanthropiens) qui vivaient
sur la côte de Benguela à Port-Alexandre. Ici, le Paléolithique inférieur correspond au passage des Néanderthaliens aux Néanthropiens alors
que le "Middle Stone Age" du Pléistocène supérieur est surtout reconnu
sur les côtes de l'Afrique du Sud. Davies (1981) situe les industries
dans le système de terrasses de l'Afrique du Sud, l'Acheuléen correspond à des stades intra-Riss ou plus ancien (Holstein), c'est-à-dire
au-delà de 200000 ans B.P. alors que le "Middle Stone Age" est localisé
entre 140000 et 125000 ans B.P.

A l'échelle des temps holocenes, les amas coquilliers culinaires
constituent des repères utiles et fréquents de l'occupation côtière par
les hommes du moment de l'approche de la mer holocène; ils sont connus
en abondance en Mauritanie, au Sénégal, en Côte d'Ivoire, au Bénin, au
Gabon, au Congo et en Angola. Par contre, ils ont pu, parfois, être à
la source de confusions avec de véritables dépôts marins, témoins di-
rects des lignes de rivage. Ceux récoltés en bordure de la lagune Ebrié
(Côte d'Ivoire) permettent de définir des populations d'abord peu lagunaires à partir de 6000 ans B.P., puis franchement lagunaires après
2500 ans B.P. (Chenorkian, 1986). En Mauritanie, l'implantation de l'homme n'intervient que vers la fin du Nouakchottien et pendant la période
lagunaire qui a suivi où les conditions de pêche étaient favorables et
la faune de savane abondante; cet exemple indique cependant un décalage
chronologique entre l'optimum écologique et le maximum de l'occupation
humaine. Les amas culinaires permettent de noter la disparition de nom-
breux biotopes en fonction des modifications du climat, du profil littoral et du régime des courants. Les industries lithiques permettent de
suivre les approches de la mer holocène entre 7 et 8000 ans B.P. à -35m
au large de Djéno (Manongo, 1985) et à 3000 ans B.P. dans les couches
saumâtres de l'embouchure de la Songololo (Giresse et Lanfranchi, 1984).
Enfin, les spectres polliniques des sédimentations côtières permettent
ainsi de souligner la présence humaine: au Nigèria, l'apparition de
l'homme vers 2800 ans B.P. est signalée grâce à la multiplication des
pollens d'*Elaeis guineensis* (Sowunmi, 1981).

LES LIGNES DE RIVAGE ET L'INTERVENTION HUMAINE

Les lignes de rivage sont le site d'aménagements qui peuvent modi-
fier leur équilibre, c'est le cas des côtes de l'Afrique du Sud où se
multiplient les implantations urbaines et les extractions de minéraux
lourds, des reculs de côte sont attribués à l'effet de pulsations transgressives, une montée de 15 cm sur une plage à pente d'un degré implique
un recul de 15m (Tinley, 1985). L'érosion ou l'engraissement du trait de
côte édifié à la fin de l'Holocène sont très contrastés selon les secteurs (progradation du SW de la Sierra Leone et du Nord de la Mauritanie). Cette évolution actuelle et future est en cours d'étude, elle

implique l'interférence de facteurs hydriques, météorologiques et océaniques avec de faibles oscillations de la composante eustatique locale; actuellement, la côte mauritanienne, au Nord du Cap Timiris montre un engraissement dû à la progression des sables dunaires, cette "régression éolienne" s'oppose à la tendance à l'érosion observée au Sud de ce Cap.

Les conséquences de l'intervention anthropique constituent des domaines de recherche en développement notamment dans le cas des équilibres de la mangrove. En Côte d'Ivoire, la dégradation récente de la mangrove est attribuée à l'action de l'homme (Frédoux, 1980); ce même recul historique est aussi enregistré sur la côte du Congo (Giresse et Lanfranchi, 1984). De même, l'interférence de la coupe des bois de chauffe et de l'extraction du sel avec la sécheresse est discutée en Afrique de l'Ouest (Paradis, 1985). Dans la baie de Sherbro (Sierra-Leone), un mouvement de retrait du rivage océanique a conduit à une exondation et à une oxydation des sols, les tannes herbacées reposent sur d'anciens sols de mangrove (Anthony et Marius, 1984-85). En Basse-Casamance, depuis 15 ans, la sécheresse provoque l'intrusion d'eaux marines vers l'amont et un fonctionnement lagunaire de la mangrove de Basse-Casamance: 70 à 80% des palétuviers ont disparu (Boivin et al., 1986).

L'édification des barrages entre aussi dans la définition de l'équilibre littoral. Dans l'arrière delta du Sénégal, le barrage de Diama a provoqué une remobilisation des sels anciens dans les zones irriguées alors qu'à l'aval, il y a risque de transformation vers un vaste bassin évaporatoire (Loyer et al., 1986).

CONCLUSIONS

Les observations relatives aux lignes de rivage du continent africain conduisent à l'analyse de facteurs qui se situent aux différentes échelles spatio-temporelles de l'évolution quaternaire ou historique ou encore de l'intervention humaine.

A l'échelle géologique de l'évolution quaternaire, les déformations tectoniques sont vérifiées dans des zones comme celles de l'Afrique du Nord à haut risque sismique et sites de tremblements de terre atteignant les populations côtières victimes de tsunamis répétés ou comme celle du rift de l'Afrique orientale où se succèdent des phases de mécanique distensive à grande échelle associée avec des petits mouvements de raccourcissement perpendiculaire et des phases de grands mouvements compressifs associés à une distension locale. Le reste de l'Afrique supposé relativement plus stable n'est pas à l'abri d'activités séismiques liées à des fracturations à grande échelle.

A l'échelle historique, plusieurs phénomènes peuvent être directement constatés voire mesurés par l'homme. L'équilibre côtier toujours précaire fait l'objet de surveillance et d'évaluation où sont enregistrées l'érosion, l'accumulation et les oscillations du niveau òcéanique. Certains phénomènes ont une cause climatique indiscutable: changements de l'hydrologie, salinisation et acidification des mangroves, accumulations éoliennes ou migrations dunaires.

Certaines actions anthropiques peuvent interférer avec les phénomènes naturels qui contrôlent la ligne de rivage: grands barrages fluvia-

tiles qui retiennent l'apport terrigène (digue d'Akosombo sur la rivière
Volta, barrage d'Inga sur le Zaïre) ou retiennent l'apport d'eau douce
dans le delta (Diama sur le Sénégal), prélèvements de sables de plage,
développement des installations portuaires qui interrompent le transit
de la dérive littorale et modifie la topographie de l'avant-côte et, par
conséquent, l'incidence des orthogonales de houle.

REFERENCES

Anthony, E.Y. and Marius, Cl., 1984–85, Géomorphologie, sédiments et
    sols de la Baie de Sherbro (Sierra Leone méridionale); Cah. ORSTOM,
    s. Pédol. Paris, XXI, 1, pp. 97–108.

Ase, L.E., 1982, Studies of shores and shore displacment on the southern
    coast of Kenya-especially in Kilifi District; Geografiska Annaler;
    63, A, pp. 303–310.

Bardet, M.G., 1974, Gisements détritiques côtiers et sous–marins du Sud-
    Ouest de l'Afrique. In "Geologie du Diamant", Mém. BRGM, 83, pp.57–
    88.

Barusseau J.P., 1984, Analyse sédimentologique des fonds marins de la
    Petite–Côte (Sénégal). Doc. Scient. Centre Recherches Océanographi-
    ques Dakar-Thiaroye, 94, 22p.

Barusseau, J.P. and Giresse P., 1987, Some mineral resources of the West
    African continental shelves related to Holocene shorelines: phospho-
    rite (Gabon, Congo), glauconite (Congo) and ilmenite (Senegal, Mau-
    ritania). In: P.G. Teleki at al (eds), Marine Minerals, Reidel Publ.
    Comp., pp.135–155.

Barusseau, J.P., Giresse P., Faure H., Lezine A.M. and Masse J.P., 1987,
    Marine sedimentary environments on some parts of the tropical and
    equatorial Atlantic margins of Africa during the Late Quaternary;
    Continental Shelf Research, Pergamon Press, 8, 1, pp.1–21.

Battistini, A., 1978, Observations sur les cordons littoraux pléistocè-
    nes et holocènes de la côte Est de Madagascar; Rev. Géographie Mada-
    gascar, XXXIII, pp. 9–37.

Battistini, A., Hinschberger, F., Morin, S. et Zogning., 1983, Le litto-
    ral du Mont Cameroun, étude géomorphologique. Rev. Géographie du
    Cameroun, IV, 1, pp. 55–72.

Bellion, Y., Hébrard, L. et Robineau, B., 1984, Sèismicité historique de
    l'Afrique de l'Ouest. Essai d'inventaire. Remarques et commentaires.
    Ass. Sénégal. Et. Quatern. Afr., Bull. Liaison, Dakar, 72–73, pp.57–
    71.

Boivin, P., Loyer, J.Y., Mougenot, B. et Sante, P., 1986. Sécheresse et
évolution des sédiments fluvio-marins au Sénégal. INQUA-ASEQUA Symposium International Dakar, "Changements globaux en Afrique", ORS-
TOM, Paris, pp. 43-38.

Brebion, P. et Ortlieb, L., 1976, Nouvelles recherches géologiques et
malacologiques sur le Quaternaire de la Province de Tarfaya (Maroc
méridional) , Géobios, 9, 5, pp. 529-55.

Brebion, P., Hoang, C.T. et Weisrock,A., 1984, Intérêt des coupes d'Aga-
dir-Port pour l'étude du Pleistocène supérieur marin du Maroc. Bull.
Mus. Natn, Paris, 4, 6, C, 2, pp. 129-151.

B.R.G.M., 1973, Recherche d'ilménite au large des côtes du Sénégal (opé-
ration Rosilda); B.R.G.M., Dept. Géol. Marine, Orléans, Repts. 73
SGN 228 MAR, 122 p.

Burke, R.C., Dessauvage, T.T.J. et Whiteman A.Y. (1972), Geological history of the Benue Valley and Adjacent Areas; African Geology, Iba-
dan, 1970, pp. 187-205.

Butzer K.W. (1972) Late Pleistocene beaches and wadi alluvia near Mersa
Alarn, Red Sea Coast, Egypt; In "Palaeoecology of Africa", ed. E.M.
van Zinderen Bakker, Balkema, Cape-Town, VI, pp.125-126.

Chenorkian, R., 1986, Caractères spécifiques et modalités d'étude des
amas coquilliers anthropiques. travaux du LAPMO, Aix-en-Provence,
32 p.

Clark, J.A. and Bloom, A.L., 1979, Hydroisostasy and Holocene emergence
of South-America; Proc. vol. Intern. Symp. On Coastal Evolution in
the Quaternary, Sao Paulo, Brazil, pp. 41-60.

Dalongeville, R. et Sanlaville, P., 1980, Etude géomorphologique de la
région Klor Eit - Marsa Arous; Travaux de la RCP 143 - Maison de
l'Orient Méditerranéen, Lyon, 15 p.

Davies, O. 1981, A reviex of Wilson's theory that the last Interglacial
ended with an ice surge, and the South African evidence therefore;
Ann. Natal Mus., 25 (1), pp. 41-59.

Diouf, M.B., Barusseau, J.P. et Giresse, P., 1986, Successions aride-
humide et formation des beach-rocks du Senegal; INQUA-ASEQUA Symposium International Dakar, "Changements globaux en Afrique", ORSTOM,
Paris, pp. 48-88.

Elouard, P., 1976. Application de la paléoécologie des Mollusques à un
problème de stratigraphie: la différenciation de deux étages du
Quaternaire marin de Mauritanie; Notes africaines, Dakar, 151, pp.
65-73.

Emery, K.O., Uchupi, E., Boivin, C.O. Phillips, J. and Simpson, E.S., 1975, Continental margin of Western Africa; Cape St-Francis (South-Africa) to Walwis Ridge (South West Africa); Bull. A.A.P.G., 59, pp. 3-59.

Fail, J.P., Montadert, L., Delteil, J.R., Valery, P., Patrait, Ph. et Schlich, R., 1970, Prolongation des zones de fracture de l'Océan Atlantique dans le Golfe de Guinée; Earth and Planetary, Sci. Letters, v. 7 pp. 413-419.

Fairbridge, R.W., 1977, Global climate change during the 13.500 b.p. Gothenburg geomagnetic excursion; Nature, 265, 5593, pp. 430-431.

Faruque, B.M., 1986, Quaternary littoral development of Ethiopia; INQUA-ASEQUA Symposium Dakar, "Changements globaux en Afrique", ORSTOM, Paris, pp. 135-136.

Faure, H. and Hebrard, L., 1977, Variations des lignes de rivage du Sénégal et en Mauritanie au cours de l'Holocène; Studia Geologica Polonica, L II, Warszawa, pp.144-157.

Faure, H., Fontes, J.C., Hebrard, L., Monteillet, Y. and Pirazzoli, P.A. 1980 a, Geoïdal change and shore-level tilt along Holocene estuaries: Senegal River Area, West Africa; Science, 210, pp. 421-423.

Faure, H., Hoang, C.T. et Lalou, C., 1980 b, Déséquilibre de l'uranium dans les calcaires coralliens et mouvements verticaux à Djibouti; In "Coll. Rift d'Asal", Djibouti, ISERST - INAG - PIRPSEV, 1p.

Frédoux, A., 1980, Evolution de la mangrove près d'Abidjan (Côte d'Ivoire) au cours des quarante derniers millénaires; Trav. et Doc. Géogr. tropicale, CEGET-CNRS, Talence, pp. 51-88.

Frédoux, A., 1983, Evolution de la végétation dans le delta de l'Agnéby (Côte d'Ivoire); Travaux et Doc. de Geogr. tropic., CEGET-CNRS, 49, Talence, pp. 117-133.

Gac, J.Y., Monteillet, J. et Faure, H., 1981, Marine shorelines in estuaries as a paleoprecipitation indicator; in "Symp. on variations in the global water budget" Oxford, U.K., 1, 2 p.

Gaven, G. et Vernier, P., 1979, Datations I_o-U de coraux et paléogéodynamique du Pléistocène moyen des îles Glorieuses (canal du Mozambique); Quaternaria, Roma, XXI, pp.45-52.

Giresse, P., 1981, Les sédimentogénèses et les morphogénèses quaternaires du plateau et de la côte du Congo en fonction du cadre structural; Bull. IFAN, Dakar, 43, A, 1-2, pp. 43-68.

150

Giresse, P., 1985, Le fer et les glauconies au large de l'embouchure du fleuve Congo; Sci. Géol., Bull. Strasbourg, 38, 4, pp.293-322.

Giresse, P., Jansen, F. Kouyoumontzakis, G. and Moguedet, G., 1981, Les fonds du plateau continental congolais et le delta sous-marin du fleuve Congo. Bilan de huit années de recherches sédimentologiques, paléontologiques, géochimiques et géophysiques; In: Milieu marin et ressources halieutiques de la République du Congo, Trav. et Doc. ORSTOM, 138, pp.17-45.

Giresse, P., Malounguila-Nganga, D. et Delibrias, G., 1984, Rythmes de la transgression et de la sédimentation holocène sur les plates-formes sous-marins du Sud du Gabon et du Congo; C.R. Acad. Sci., Paris, 229, II, 7, pp.327-330.

Giresse, P. et Lanfranchi, R., 1984, Les climats et les océans de la région congolaise. Bilans selon les échelles et les méthodes de l'observation; Palaeoecology of Africa, J.A. Coetze and E.M. Van Zinderen Bakker Sr éd, Rotterdam, 16, pp. 77-88.

Grant, N.K., 1971, South Atlantic Benue Trough and Gulf of Guinea triple junction; Geol. Soc. America Bull., 82, pp.2295-2298.

King, B.C., 1970, Vulcanicity and rift tectonics in East Africa, in: T. N. Clifford and I.G. Grass éd., African Magmatism and Tectonics, Edinburgh, pp. 263-283.

Kudrass, H.R., 1987, Sedimentary models to estimate the heavy-mineral potentiel of shelf sediments; in : P.G. Teleki et al. (eds), Marine Minerals, Reidel Publ. Comp., pp. 39-56.

Lagaaij, R., 1973, Shallow-water Bryozoa from Deep-sea of the Principe channel, Gulf of Guinea; in : Living and Fossil Bryozoa, London Acad. Press, Larwood G.D. ed, pp. 139-151.

Le Pichon, X. and Hayes, D.E., 1971, Marginal offsets, fracture zone and the early opening of the South Atlantic; Journ. of Geophys. Research, 76, 26, pp. 6283-6293.

Loyer, J.Y., Mougenot, B. et Zante, P., 1986, Changements récents induits par l'intervention humaine sur le sol de la basse-vallée du fleuve Sénégal; INQUA-ASEQUA Symposium International, Dakar, "Changements globaux en Afrique", ORSTOM, Paris, pp. 43-48.

Mc Master, R.L., Milliman, J.D. and Ashraf, A., 1971, Continental shelf and upper slope sediments off Portuguese Guinea and Sierra-Leone, West Africa; Journal of Sedimentary Petrology, 41, pp. 150-158.

Manongo, L., 1985, Le gisement sous-marin des phosphates de Djéno. Nouvelles observations. D.E.A. Univ. Toulouse-Perpignan, 56 p.

Martin, L., 1973, Morphologie, sédimentologie et paléogéographie au Quaternaire récent du plateau continental ivoirien; Th. Doct. ès Sc. Univ. Paris VI, 340 p.

Monteillet, J., Faure, H., Pirazzoli, P.A., et Ravisé, A., 1981, L'invasion saline du Ferlo (Sénégal) à l'Holocène supérieur (1900 B.P.); in Palaeoecology of Africa and the surrounding islands, J.A. Coetze and E.M. Wan Zinderen Bakker Sr. ed, Rotterdam, 13, pp. 205-216.

Morner, N.A., 1976, Eustasy and geoïd changes; Journ. of Geol., 84, pp. 123-151.

Paradis, G., 1986, Rôle de l'homme dans les changements du paysage tropical: les mangroves ouest-africaïnes; INQUA-ASEQUA Symposium Intern. Dakar "Changements globaux en Afrique, ORSTOM, Paris, pp. 349-351.

Peypouquet, J.P., 1977, Les ostracodes indicateurs paléoclimatiques et paléogéographiques du Quaternaire terminal (Holocène) sur le plateau continental sénégalais. In: 6th Intern. Symp. on Ecology and Zoogeography of Recent and Fossil Ostracoda, tt. Loffer et D.W.W. Jung Publ., Saafeden, Salzburg, pp. 369-394.

Pinson-Mouillot, J., 1980, Les environnements sédimentaires actuels et quaternaires du plateau continental sénégalais (Nord de la presqu'-île du Cap Vert); Thèse de Doct. de sp. Bordeaux, 106 p.

Ramos, M., 1981, As exavacoes de Capangombe eo problema da M.SA., SW de Angola; Leba, 4, pp. 43-52.

Ramos, M., 1982, Le Paléolithique du Sud-Ouest de l'Angola, vue d'ensemble; Leba, 4, pp. 43-52.

Siedner, R. and Miller, J., 1968, K/AR age determination on basaltic rocks from South-West Africa and their bearing on continental drift; Earth Plan. Sci. Let. 4, pp. 1451.

Soares de Carvalho, G., 1961, Alguna problemas dos terracos quaternarias de littòral de Angola; Bol. Serv. geol min Angola, 2, pp.5-15.

Sowunmi, M.A., 1981, Late Quaternary environmental changes in Nigeria; Pollens et Spores, XXIII, I, pp. 125-148.

Stearns, C.E., 1978, Pliocene-Pleistocene emergence of the Morrocan Meseta; Geol. Soc. Amer. Bull., 89, pp. 1630-1644.

Tastet, J.P., 1979, L'Holocène du littoral septentrional du Golfe de Guinée; Proc. of 1978 Intern. Symp. on coastal evolution in the Quaternary, Sao Paulo, Brazil, pp. 588-606.

Thomson, J, and Walton, A., 1972, Redetermination of Chronology of Aldabra Atoll by $^{230}TH/^{234}U$ dating, Nature, 240, pp. 145-146.

Tinley, K.L., 1985, Coastal dunes of South-Africa; South African National Sc. Programmes, Pretoria, Rep., 109, 300 p.

Thomson, J, and Walton, A., 1972, Redetermination of Chronology of Aldabra Atoll by $^{230}TH/^{234}U$ dating, Nature, 240, pp. 145-146.

Unesco, 1986, Principles of geological mapping of marine sediments (with special reference to the African continental margin), Unesco rep. in Marine Science, 37, 101 p.

Veenstra, H.J. and Winkelmolen, A.M., 1976, Size shape and density sorting around two barrier islands along the north coast of Holland; Geol. Mijnbouw, 55, pp. 87-104.

Weisrock, A., 1980, The littoral deposits of the Saharian atlantic coast since 150.000 years; Palaeoecology of Africa, J.A. Coetze and E.M. Van Zinderen Bakker ed., Rotterdam, 12, pp. 277-287.

Weisrock, A., 1981, Variations du niveau de l'océan et morphologie littorale du Haut-Atlas (Maroc) depuis 100.000 ans; Oceanis, 7, 4, pp. 481-487.

Weisrock, A. Delibrias, G. Rognon, P. and Coude-Gaussen, G., 1985, Variations climatiques et morphogénèses au Maroc atlantique (30-33°N) à la limite Pleistocène-Holocène; Bull. Soc. Géol. France, 8, 1, 4, pp. 565-569.

RECENT SEA-LEVEL CHANGES IN THE NORTH ATLANTIC

P. A. Pirazzoli
16, rue de la Grange Batelière
75009 Paris
France

ABSTRACT. Most of the tide-gauge stations still active, for which long series of almost continuous records are available, are located on both sides of the North Atlantic. Only records starting earlier than 1925 have been considered in this paper : 17 stations on the American side, and 58 on the European one. In addition records in the Bermudas starting in 1933 have been used for regional comparisons. All the stations indicate secular trends of relative sea-level variation which change from place to place. The data have been grouped in 5-yr periods, but the secular trends have been left out of the data in order to remove crustal movements.

   Short term (5 to 20-yr) regional oscillations appear. Average 5-10 yr changes in sea level on the American coasts (where a north-south gradient is frequent) differ from those on the European ones. They often differ too between the American and the Bermuda coasts. This suggests that changes in the sea-surface topography across the Gulf Stream and in the sea-level slope occur in the North Atlantic, making regional sea levels change from one area to the other, with average differences in level reaching as much as one decimetre over 5-yr periods. Also on the European coasts changes in sea level often differ from one region to the other.

   After an estimation of the eustatic component obscured by the local secular trends, the conclusion is reached that the average sea level, which shows faster rates of change on the American coasts, rose on both sides of the Atlantic from 1920 to 1950, and then remained stable or dropped slightly between 1950 and 1980, in spite of the increasing CO_2 atmospheric content. In Europe, the sea-level rise during the last century that can be ascribed to eustatic changes is only a few centimetres.

1. INTRODUCTION

Recent studies (NAS, 1983 ; EPA, 1983 ; MacCracken & Luther, 1985 ; Bolin et al., 1986) have shown that the atmospheric concentration of carbon dioxide and of several gases produced by industrial activi-

153

D. B. Scott et al. (eds.), Late Quaternary Sea-Level Correlation and Applications, 153–167.
© 1989 by Kluwer Academic Publishers.

ties has been increasing since the end of the 18th century, especially during the last few decades. In particular CO_2 concentration in the atmosphere has increased from an estimated preindustrial level of 260-270 ppm, deduced from the analysis of air bubbles entrapped in Antarctic ice cores (Reynaud & Barnola, 1985), to 315 ppm in 1958 and 345 ppm in 1985 in Hawaii, and it has been predicted that this rise will continue or even accelerate. This increase is expected to create a so-called "greenhouse effect" and produce an increase in the atmospheric and oceanic temperatures, a melting of glaciers and land-based ice sheets, and a rise of sea level.

Several authors (see Pirazzoli (1986) for references) have assumed that if tide-gauge data are supplied by a sufficient or "well-distributed" number of stations, crustal movements would be compensated and average changes in the relative sea level (RSL) would depend mainly on variations of the water volume. Accordingly, a "global" sea-level rise of 1.0 to 1.5 mm/yr during the last century has been inferred. This rise, which has been ascribed to glacial melting (glacio-eustasy) and/or to steric (thermal) effects (Gornitz et al., 1982), would be progressing or even accelerating at present. It has been shown, however, that the average of all the trends indicated by tide-gauge stations may be biased towards a RSL rise by systematic down-warping of coastlines (Pirazzoli, 1986).

In this paper a new approach is proposed and it is shown that the available sea-level data do not justify overalarmist interpretations, since the average sea level has been dropping rather than rising in the North Atlantic during the last three decades.

Fig. 1.   Location of the tide-gauge stations studied.

A : East coasts of North America. (1) Pointe-au-Père ; (2) Charlottetown ; (3) Halifax ; (4) Portland ; (5) Boston ; (6) New York ; (7) Atlantic City ; (8) Philadelphia ; (9) Lewes ; (10) Baltimore ; (11) Charleston ; (12) Fernandina ; (13) Key West ; (14) Cedar Keys ; (15) Pensacola ; (16) Galveston ; (17) Cristobal.

B : Coasts of Europe. (1) Bergen ; (2) Stavanger ; (3) Oslo ; (4) Strömstad ; (5) Smögen ; (6) Hirsthals (7) Esbjerg ; (8) Delfzijl ; (9) Terschelling ; (10) Harlingen ; (11) Den Helder ; (12) Ijmuiden ; (13) Hoek van Holland ; (14) Hellevoetsluis ; (15) Brouwershaven ; (16) Zierikzee ; (17) Vlissingen ; (18) Aberdeen II ; (19) North Shields ; (20) Sheerness ; (21) Newlyn ; (22) Liverpool ; (23) Brest ; (24) Cascais ; (25) Lagos ; (26) Göteborg ; (27) Varberg ; (28) Ystad ; (29) Kungholmsfort ; (30) Olands Norra Udde ; (31) Landsort ; (32) Stockholm ; (33) Björn ; (34) Draghällan ; (35) Ratan ; (36) Furuögrund ; (37) Kemi ; (38) Oulu ; (39) Jacobstad ; (40) Vaasa ; (41) Mäntyluoto ; (42) Turku ; (43) Hangö ; (44) Helsinki ; (45) Gedser ; (46) Köbenhavn ; (47) Hornbaek ; (48) Korsör ; (49) Slipshavn ; (50) Fredericia ; (51) Aarhus ; (52) Frederishavn ; (53) Alicante ; (54) Marseille ; (55) Genova ; (56) Venezia ; (57) Trieste ; (58) Port Tuapse.

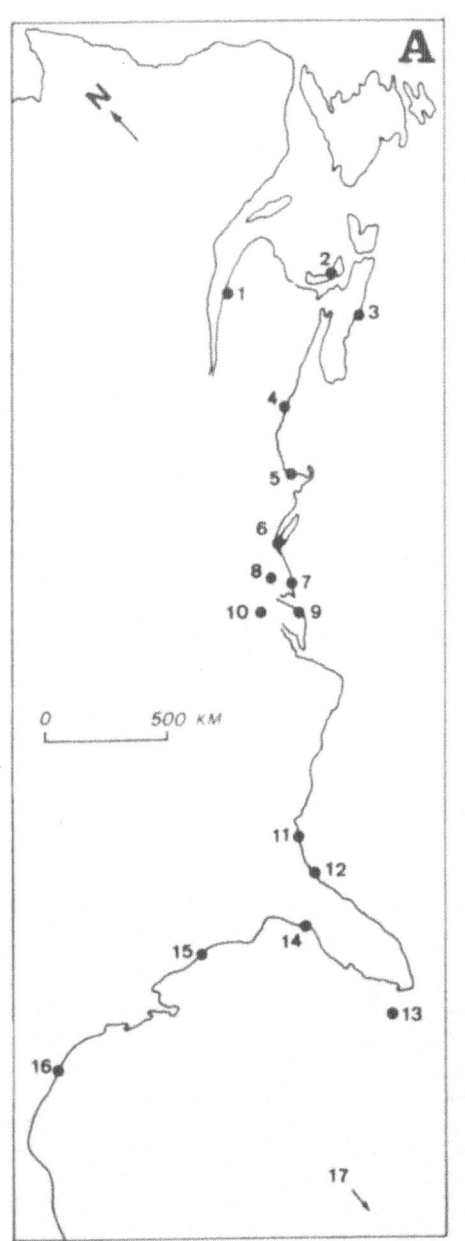
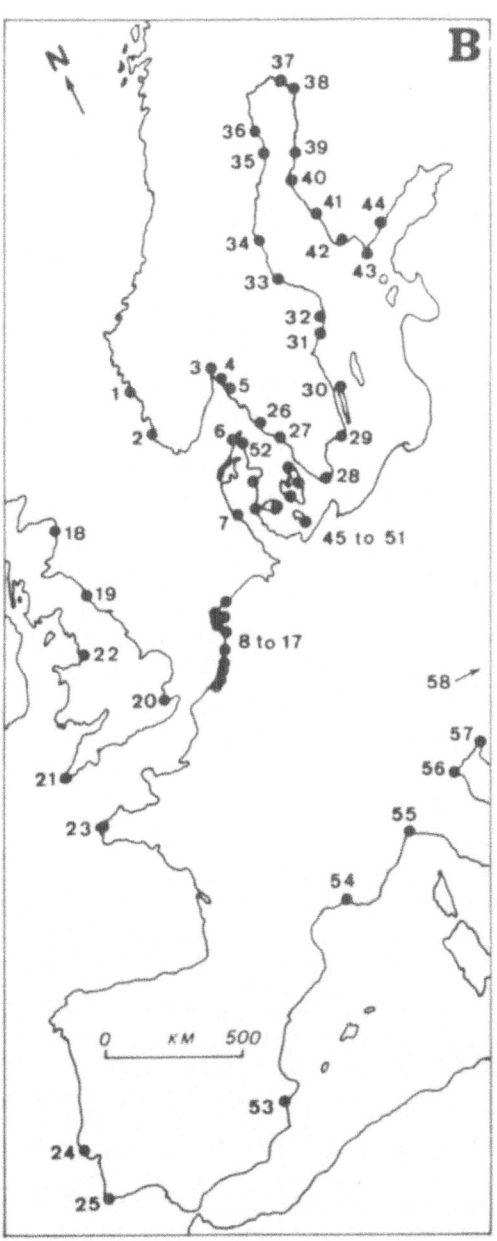

2. THE DATA USED

Although records of some 1178 stations can be provided by the Permanent Service for Mean Sea Level (1), or found in the literature, only 229 stations have records long enough to enable the computation of acceptable secular trends (Hicks et al., 1983; Barnett, 1984; Pirazzoli, 1986). Among these 229 stations, 118 are located on both sides of the North Atlantic, but only 75 stations (17 on the American side and 58 on the European one), which have records starting earlier than 1925 and are still current, have been used in the present study (Fig. 1). In addition records for the Bermudas starting in 1933 have been used for regional comparisons.

3. DISTRIBUTION OF SECULAR RSL TRENDS

The rates of secular RSL change at each station, compiled from Hicks et al. (1983) and Pirazzoli (1986), are reported graphically in Fig. 2.

On the American side, with the exception of station 1 (Pointe-au-Père), which is situated in an area of glacio-isostatic uplift, most of the tide gauges indicate a long term RSL rise, at rates varying from about 2 to 4 mm/yr and reaching as much as 6.3 mm/yr at Galveston, where anthropogenic effects are predominant.

A submergence trend along these coasts is predicted by global isostatic models (Walcott, 1972; Clark et al., 1978). However the RSL rise observed is higher than the rates deduced from geologic evidence for the late Holocene (e.g 2.3 mm/yr in Boston and 2.8 mm/yr in New York from tide-gauge records, whereas the geologic trend since 2100 yr BP in the Barnstable Marshes is about 1 mm/yr (Redfield & Rubin, 1962); 3.0 mm/yr from tide gauges at Lewes, whereas the geologic trend in coastal Delaware since 2000 yr BP is about 1.25 mm/yr (Belknap & Kraft, 1977); 2.0 mm/yr from tide gauges at Key West, whereas the geologic trend in the nearby Everglades has been about 0.25 mm/yr since 1000 yr BP (Scholl & Stuiver, 1967)). The possible causes of the recent accelerated rise are still subject to speculation, since a "global" eustatic change of 1.0 to 1.5 mm/yr does not suffice to explain the present rates of RSL rise in many cases.

On the European coasts the overall pattern is clearly that foreseen by global isostatic models: a dome-shaped uplift centred in the Gulf of Bothnia, surrounded by a wide belt of subsidence (Emery & Aubrey, 1985). If those stations indicating a RSL rise (2, 7 to 25, 28, 45 to 58) are considered separately, they indicate an average rate of rise of 1.26 mm/yr, i.e. of the same order as the "global" eustatic rise claimed by several authors. However, for rheological reasons, a subsidence movement in this area must necessarily exist, to balance the uplift movements in Fennoscandia. This means that the recent eustatic rise in Europe must certainly be less than 1.26 mm/yr.

----

(1) Address: P.S.M.S.L., P.O.L., Bidston Observatory, Birkenhead, Merseyside, U.K.

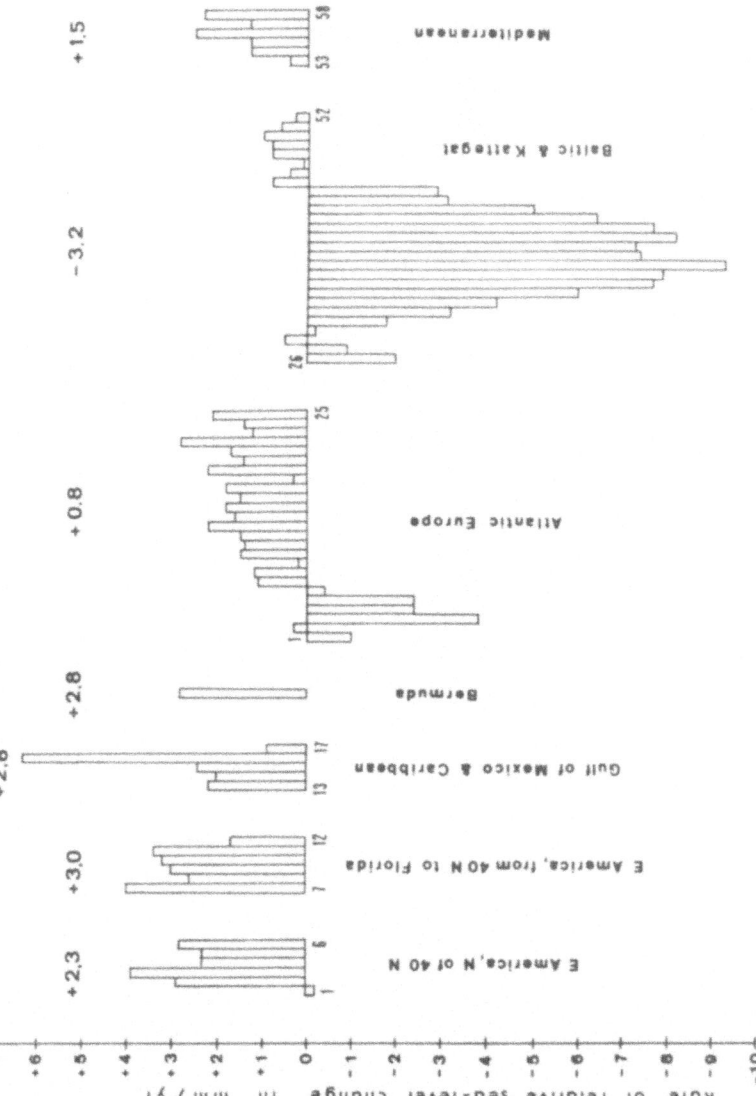

Fig. 2. Local secular trends of RSL change at the tide-gauge stations studied. The numbers near the zero line in the graphs correspond to the location numbers in Fig. 1. The numbers above the graphs are average rates of RSL change in each area (+ = sea-level rise; − = sea-level drop).

The average regional RSL changes prior to 1980 are reported graphically in Fig. 3. The changes clearly vary from one area to the other and no general trend predominates. Indeed the determination of the present eustatic trend remains at this stage an unanswered, challenging problem, even in the area of the world where the network of tide gauges is most dense and the records available longest.

4. REGIONAL 5-YEAR MEAN SEA-LEVEL CHANGES

The tide-gauge records available at each station were grouped in 5-year periods and for each period the 5-year average sea level was calculated and assigned to the middle year. In order to filter crustal movements, the local secular trend was subtracted at each station, although it may include an eustatic component (this point is discussed below). Lastly the values obtained were grouped in the seven areas defined previously and in each area the value of the average 5-year sea-level

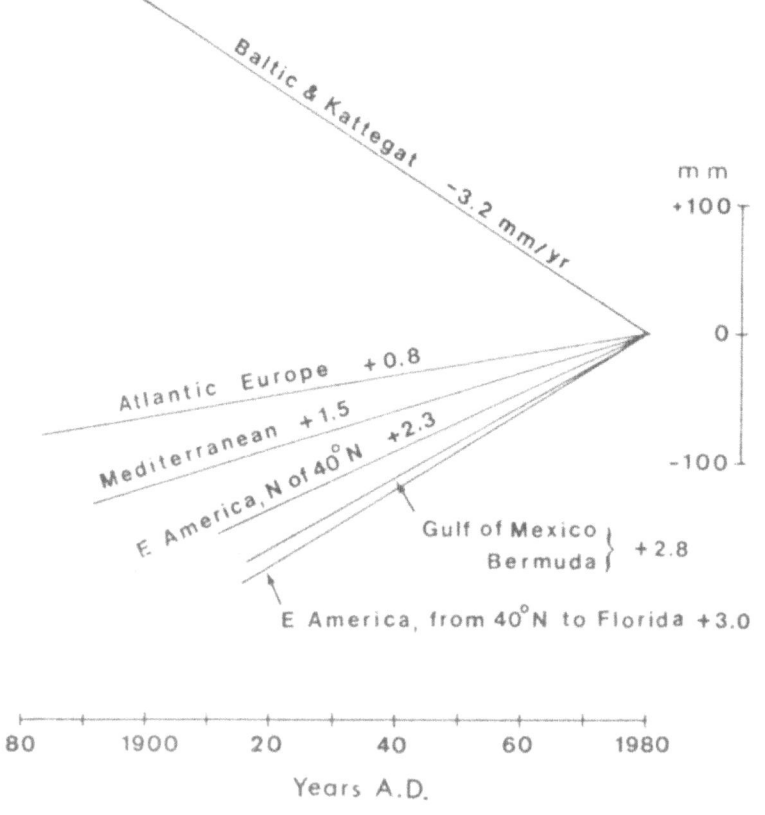

Fig. 3 Average RSL changes during the last century in the areas studied.

Table I

Average 5-year sea-level changes in the North Atlantic in relation to the preceding 5-year period (in mm; + = sea-level rise). Secular trends have been subtracted at each station.

Periods of time	America north of 40° N	America from 40° N to Florida	Gulf of Mexico & Caribbean	Atlantic Europe	Baltic & Kattegat	Mediter- ranean	East N America	Bermuda	Europe
	1	2	3	4	5	6	1 to 3		4 to 6
1976-1980	- 10.4	- 64.2	- 52.0	+ 8.9	- 44.6	+ 26.8	- 42.2	+ 44.0	- 3.0
1971-1975	+ 0.5	+ 47.5	+ 57.2	- 38.3	+ 8.4	- 41.2	+ 35.1	+ 25.0	- 23.7
1966-1970	+ 4.8	+ 5.2	- 2.0	+ 17.9	- 12.9	+ 10.7	+ 2.7	- 75.0	+ 5.2
1961-1965	- 24.8	- 24.7	- 24.0	+ 10.1	+ 36.6	+ 5.2	- 24.5	- 15.0	+ 17.3
1956-1960	- 15.8	- 1.2	+ 13.4	- 16.1	- 20.7	- 5.7	- 1.2	+ 42.0	- 14.2
1951-1955	+ 17.5	- 13.8	- 50.4	- 3.9	- 15.0	+ 35.8	- 15.6	+ 14.0	+ 5.6
1946-1950	- 7.5	+ 16.0	+ 27.8	+ 15.7	+ 14.7	- 18.2	+ 12.1	+ 8.0	+ 4.0
1941-1945	+ 11.4	- 9.0	+ 26.0	- 1.1	+ 33.6	- 23.8	+ 9.5		+ 2.9
1936-1940	+ 8.0	+ 21.2	+ 16.2	- 1.6	- 33.0	+ 32.2	+ 15.1		- 0.8
1931-1935	+ 26.0	+ 23.0	- 15.5	- 8.4	- 0.5	+ 4.2	+ 11.2		- 1.6
1926-1930	- 11.3	- 19.7	- 3.2	+ 6.6	- 27.4	+ 0.8	- 11.4		- 6.7
1921-1925		- 15.3	+ 9.7	+ 6.3	+ 46.7	- 31.4	(- 2.8)		+ 7.2
1916-1920		+ 3.2	- 15.5	- 24.4	- 35.8	+ 10.7			- 16.5
1911-1915				+ 29.3	+ 28.3	+ 8.7			+ 22.1
1906-1910				+ 3.0	- 5.1	- 32.7			- 11.6
1901-1905				- 1.8	- 19.2	+ 19.7			- 0.4
1896-1900				- 10.7	+ 21.5	- 9.5			+ 0.4
1891-1895				+ 20.6	+ 15.1	+ 4.2			+ 13.3
1886-1890				- 26.2		- 9.7			
1881-1885				- 38.0					
1876-1880				+ 42.1					
1871-1875				- 13.1					
standard deviations	14.6	26.4	29.8	16.6	26.6	21.3	20.4	38.1	15.4

change was computed in relation to the preceding 5-year period. The results are shown in Table 1, where the last columns give average values obtained algebraically from the preceding ones and the last line, the standard deviations for each regional series of 5-yr values.

Attention must be drawn to certain points. On the American side of the Atlantic the sea-level change shows a north-south gradient half the time. The greatest average changes in sea level are usually

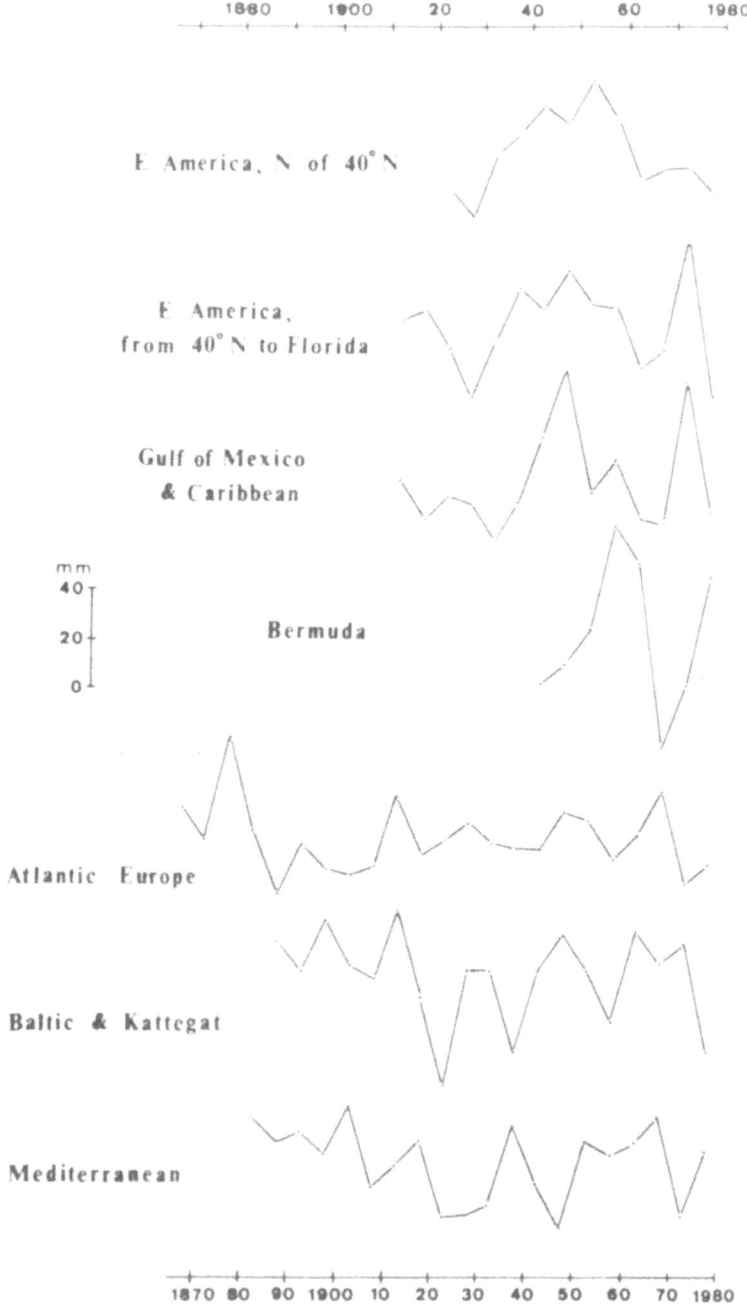

Fig. 4.   Average 5-year sea-level variations in the areas studied
           after subtraction of the local secular trend at each station.

Fig. 5   Regional average 5-year sea-level variations after
subtraction of the local secular trends.

found at the lowest latitudes. This seems to indicate that the changes
in sea level observed depend on climatic and oceanic factors, rather
than on glacio-eustasy.

The direction of the overall trend of the 5-year sea-level
change on the American side (areas 1 to 3) is often opposite to that
of the contemporaneous changes in the Bermudas. The low correlation
coefficient between these two data series (-0.12) shows that they
are almost unrelated to each other. It can be inferred that changes
in the ocean-surface profile across the Gulf Stream are a very common
phenomenon.

On the Atlantic coasts of Europe (area 4), although the
average correlation between stations is generally weak (Woodworth,
1986), the direction of sea-level change is often contrary (correlation
coefficient : -0.55) to that occurring at the same time on the American
side (areas 1 to 3). The Mediterranean and the Baltic both have the
same low correlation coefficient (0.26) with the nearby Atlantic,
although their individual behaviour is often opposite (correlation
coefficient between the Mediterranean and the Baltic : -0.58). On
the whole the sea level does not rise or drop simultaneously everywhere
in the North Atlantic ; but changes from one area to the other, with
average differences in level reaching as much as one decimetre over
5-year periods.

The oscillations observed may be associated, although not
exclusively, with modifications of barometric pressure and winds.
In the U.K., for example, local winds and air pressure are responsible
for a large fraction of sea-level annual variability (Woodworth, 1987).
Similarly, on the American side, meteorology (particularly wind stress)
plays a major role (Blaha, 1984). Moreover these fluctuations are
not correlated across the Atlantic (Thompson, 1986).

The changes manifest in Table 1 can be seen from the time
series of Fig. 4 and 5. A 35-year wave about 5 cm high is manifest
on the east coast of North America, followed by a 10-year wave of
similar amplitude, but having developed south of latitude 40° N only.

These movements, as well as the strong sea-level oscillations in Bermuda, possibly depend on changes in the Gulf Stream, where satellites have shown a high sea-level variability (Cheney et al., 1984). In Europe, on the other hand, the 5-year sea-level movement appears much more erratic and limited.

5. DISCUSSION AND CONCLUSIONS

The question arises as to whether part of the eustatic change may be obscured by the secular trends which have been subtracted from the sea-level data.

A global eustatic fluctuation is expected to be correlated with the mean surface air temperature, changes in which would cause glacial melting or retreat, and steric effects in the ocean. The mean surface temperature has probably increased about 1° C since the "Little Ice Age", which is considered to have lasted from the end of the 16th century to the middle of the 19th century. Most of this increase seems to have occurred between the 1880's and the 1940's, when a retreat of the ice front is reported in many mid-latitude glaciers. From 1950 to 1980 the mean surface temperature diminished slightly in the Northern Hemisphere according to Yamamoto and Hoshiai (1980), or was more or less stable on a world scale (Jones et al., 1986) ; the mid-latitude glaciers have been almost stable or have started advancing again (CGI, 1986). Accordingly a slight eustatic rise can be expected between 1880 and 1950, and a slight sea-level drop, or a period of sea-level stability, after 1950.

An estimation of the eustatic sea-level changes can be attempted at two stations with exceptionally long records and relatively limited vertical land movements : Amsterdam and Brest. In Amsterdam, from 1682 to 1930, the RSL rise was about 17 cm (Van Veen, 1954). However, if a local subsidence rate of 0.4 mm/yr is taken into account, four stages appear : (1) eustatic stability from 1682 to 1740; (2) eustatic fall at a mean rate of 0.25 mm/yr from 1740 to 1830 ; (3) again eustatic stability from 1820 to 1840 ; (4) lastly, eustatic rise at a mean rate of 0.9 mm/yr from 1840 to 1930 (Mörner, 1973). According to the same author a comparison with tide-gauge records from the Baltic shows that the eustatic rise slowed down about 1930 and ended about 1950.

In Brest, where almost continuous records are available since 1807, the overall trend is towards a rise of 0.8 mm/yr, but an acceleration can be observed from 0.3 mm/yr until about the end of the last century, to 1.2 mm/yr after that time (Pirazzoli, 1986). From 1950 to 1980 the average rate of RSL rise is less than 0.2 mm/yr and there has been a distinct drop since the end of the 1960's, as already noted by Woodworth (1987) in the U.K. and in Atlantic Europe.

The above data for both areas suggest that an eustatic rise, at an average rate of about 0.9 mm/yr, occurred between the second half of the 19th century and the middle of the 20th century on the Atlantic coasts of Europe. Unfortunately no other stations can provide

Table II

Average sea-level changes and rates of change in the North Atlantic from 1920 to 1980

	1920 to 1950			1950 to 1980		
	sea-level change	rate	corrected eustatic rate	sea-level change	rate	corrected eustatic rate
	(mm)	(mm/yr)	(mm/yr)	(mm)	(mm/yr)	(mm/yr)
	1	2	3	4	5	6
Europe	+ 5.0	+0.16	+0.75	-12.8	-0.4	+0.2
East N America	+33.7	+1.12	+1.6 ?	-45.7	-1.5	-1.0 ?

Note : local secular trends have been subtracted from the data in columns 1, 2, 4 and 5. The estimated eustatic component obscured by this subtraction has been added in columns 3 and 6. + = sea-level rise; - = sea-level drop.

records long and continuous enough to make a similar estimation possible in other areas.

The periods of time considered for the computation of the secular trends used in this paper are not the same everywhere but take place in Europe within the period 1880 to 1950 (66 %), before 1880 (4 %) or after 1950 (30 %). On the American coasts 56 % of the tide-gauge records used date from 1880 to 1950 and 44 % from after 1950. Assuming that the eustatic sea-level has been almost stable since 1950, it can be estimated that the secular trends which have been subtracted from the sea-level data obscure, in Europe, a eustatic component corresponding to a rise of about 0.6 mm/yr (66 % of 0.9 mm/yr). When

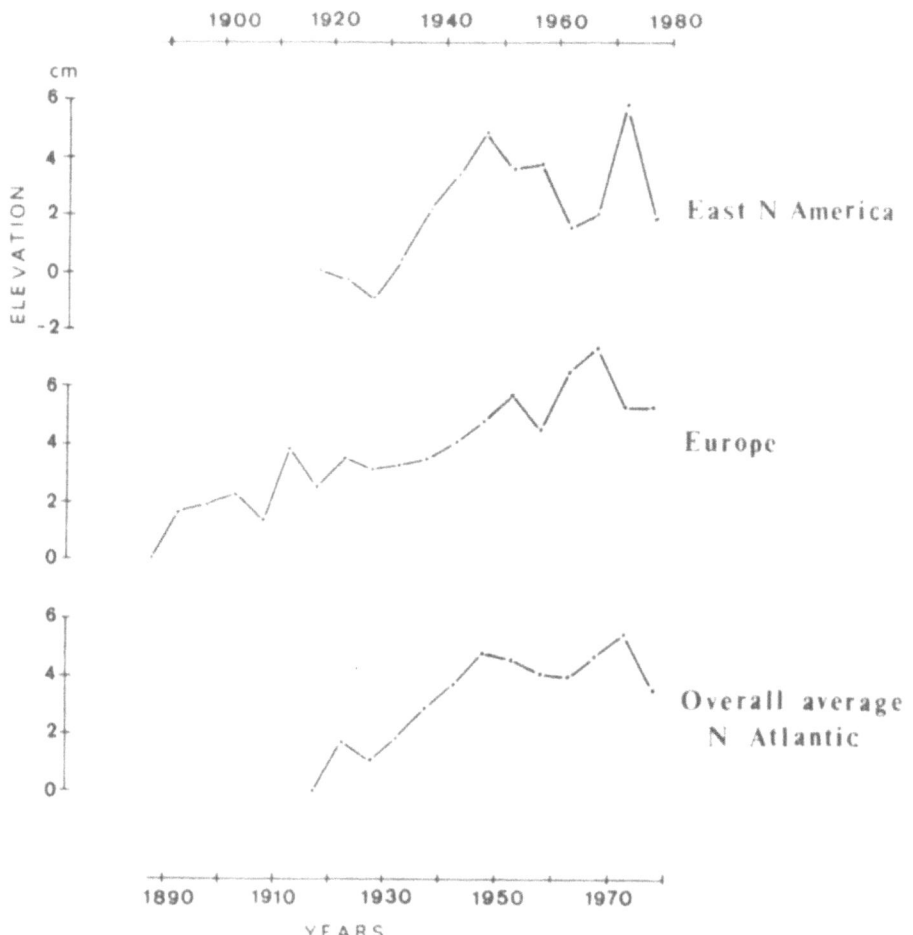

Fig. 6   Regional average 5-yr sea-level variations in the North Atlantic. The local secular trends have been subtracted from the data, but the eustatic component included in the secular trends has been added.

this value is added to the European data, the total average rate of sea-level rise from 1920 to 1950 becomes 0.75 mm/yr (Table 2). After 1950, the assumption that the eustatic sea level has remained almost stable in Europe is confirmed by the very limited rise (0.2 mm/yr) which is obtained when the value 0.6 mm/yr is added to the data.

In the absence of direct estimations of the eustatic changes on the American coasts, it can be assumed tentatively that the eustatic rise was the same here as in Europe, e.g. 0.9 mm/yr from 1880 to 1950. Accordingly the component of this rise obscured by the local secular trends would be about 0.5 mm/yr (54 % of 0.9 mm/yr) and the total average rate of sea-level rise from 1920 to 1950 about 1.6 mm/yr. After 1950 a marked sea-level drop is apparent on the American coasts, more so if the value 0.5 mm/yr is added to the data. These results show that the assumption of a similar eustatic movement on both sides of the North Atlantic Ocean is not correct and this precludes more precise estimations on the coasts of North America.

In conclusion, (Fig. 6) the average regional sea level rose on both sides of the North Atlantic from 1920 to 1950, but the rates of change vary distinctly from one area to the other and have been much faster on the American side than on the European one, possibly in relation to changes in the Gulf Stream and in the Atlantic circulation.

If local crustal movements are filtered, the average sea-level rise since the end of the last century is only a few centimetres. From 1950 to 1980 the average sea level has been almost stable on the coasts of Europe, dropping clearly on the coasts of America. This means that no distinct relation can yet be detected between the increase of CO_2 and of other so-called "greenhouse" gases in the atmosphere and sea level.

6. ACKNOWLEGMENTS

This paper is a contribution to the IGCP Project 200. The author is indebted to R.E. Thomson (Institute of Ocean Sciences, Sydney, Canada) and to P. Woodworth (P.O.L., Bidston Observatory, U.K.) for their friendly criticism and many helpful comments and ideas, to Ms. M. Delahaye (Institut Océanographique, Paris) for revising the English text and to Ms. V. Collard and Ms. D. Maraine (Intergéo, Paris) for typing and page-setting the manuscript.

7. REFERENCES

Barnett T.P., 1984. The estimation of "global" sea level change: a problem of uniqueness. J. Geophys. Res., 89, C5 : 7980-7988.
Belknap D.F. & Kraft J.C., 1977. Holocene relative sea-level changes and coastal stratigraphic units on the northwest flank of the Baltimore Canyon Trough geosyncline. J. Sedim. Petr., 47, 2: 610-629.

Blaha J.P., 1984. Fluctuations of monthly sea level as related to the intensity of the Gulf Stream from Key West to Norfolk. J. Geophys. Res., **85**, C5: 8033-8042.

Bollin B., Döös B., Jager J. & Warrick R.A. (Eds.), 1986. The greenhouse effect, climatic change and ecosystems. SCOPE 29. Chichester, J. Wiley & Sons.

Cheney R.E., Marsh J.G. & Martin T.V., 1984. Applications of satellite altimetry to oceanography and geophysics. Marine Geophys. Res., **7**: 17-32.

Clark J.A., Farrell W.E. & Peltier W.R., 1978. Global changes in postglacial sea level: a numerical calculation. Quatern. Res., **9**: 265-287.

Comitato Glaciologico Italiano (CGI), 1986. Atti del 5° Convegno Glaciologico Italiano. Geogr. Fis. Din. Quatern., **8** (2), 1985: 65-214.

Environmental Protection Agency (EPA), 1983. Can we delay a greenhouse warming ? EPA 230-09.007. Washington D.C.

Emery K.O. & Aubrey D.G., 1985. Glacial rebound and relative sea levels in Europe from tide-gauge records. Tectonophysics, **120**: 239-255.

Gornitz V., Lebedeff S. & Hansen J., 1982. Global sea-level trend in the past century. Science, **215**: 1611-1614.

Hicks S.D., Debaugh H.A., Jr. & Hickman L.E., 1983. Sea level variations for the United States 1855-1980. Rockville, Maryland: NOAA National Ocean Service, 170 p.

Jones P.D., Wigley T.M.L. & Wright P.B., 1986. Global temperature variations between 1861 and 1984. Nature, **322**, 6078: 430-434.

MacCracken M.C. & Luther F.M. (Eds.), 1985. Projecting the climatic effects of increasing carbon dioxide. DOE/ER-0235: US Department of Energy, Washington D.C.

Mörner N.A., 1973. Eustatic changes during the last 300 years. Palaeogeogr., Palaeoclim., Palaeoecol., **13**: 1-14.

National Academy of Sciences (NAS), 1983. Changing climate. Washington D.C., National Academy Press.

Pirazzoli P.A., 1986. Secular trends of relative sea-level (RSL) changes indicated by tide-gauge records. J. Coast. Res., S.I.1: 1-26.

Redfield A.C. & Rubin M., 1962. The age of salt marsh peat and its relation to recent changes in sea level at Barnstable, Massachussetts. Proc. Nat. Acad. Sci. (USA), **48**, 10: 1728-1735.

Raynaud D. & Barnola J.M., 1985. An Antarctic ice core reveals atmospheric CO_2 variations over the past few centuries. Nature, **315**: 309-311.

Scholl D.W. & Stuiver M., 1967. Recent submergence of southern Florida. Geol. Soc. Am. Bull., **78**: 1195-1197.

Thompson K.R., 1986. North Atlantic sea level and circulation. Geophys. J.R. astr. Soc., **87**: 15-32.

Van Veen J., 1954. Tide-gauges, subsidence gauges and flood-stones in the Netherlands. Geol. Mijn., **16**: 214-219.

Walcott R.I., 1972. Past sea levels, eustasy and deformation of the Earth. Quatern. Res., **2**: 1-14.

Woodworth P.L., 1986. A global sea-level network: how many gauges are enough ? Tropical Ocean-Atmosphere Newsletter, October: 3-6.

Woodworth P.L., 1987. Trends in U.K. mean sea level. Marine Geodesy,
    11: 57-87.
Yamamoto R. & Hoshiai M., 1980. Fluctuations of the Northern Hemi-
    sphere mean surface air temperature during recent 100 years,
    estimated by optimum interpolation. J. Meteor. Soc. Japan, 58,
    3: 187-193.

LATE QUATERNARY SHORELINES IN INDIA

Helmut Brückner
Department of Geography
University of Düsseldorf
Universitätsstrasse 1
D-4000 Düsseldorf 1
Federal Republic of Germany

ABSTRACT. Littoral lowlands in India were examined to find indicators for higher sea levels, date the sediments (with the ^{14}C, [230]Th/[234]U and Electron Spin Resonance dating techniques) and reconstruct the geomorphologic and tectonic evolution during the late Quaternary. – On Kathiawar Peninsula the miliolite problem (marine versus eolian genesis) could be solved. The inland occurrences of these organogenic grainstones are of eolian origin. In the coastal belt miliolites up to +4 m above mean sea level were deposited by the sea during the last interglacial transgression. This is confirmed by the dating of other marine accumulation terraces up to this altitude which are also of stage 5e of the oxygenisotope record (c. 125 ka). The Holocene transgression reached its maximum around 5,000-6,000 BP when sea level was c. 1 m higher (^{14}C-dated marshy soils and marine sands). The Konkan Coast between Bombay and Goa is a ria-type coast. So far no Pleistocene marine deposits have been detected. Cross-sections of the lower river courses and creeks also indicate subsidence. Here, the Holocene transgression reached its maximum c. 2,000-2,800 BP, was at most 1 m higher and deposited beach ridges. On the Tamilnadu side of southern India, between Cape Comorin and Rameswaram, the 125 ka-strandline is found occurring not higher than 4 m above present sea level, which indicates subsidence or stability of the coast rather than emergence. The Holocene transgression reached up to +1 m. The Kerala side is devoid of Pleistocene marine deposits above present sea level, and even the Holocene ones are drowned. – It is worth noting that in Peninsular India Pleistocene marine deposits older than stage 5 are missing. It seems that for geologically old continents other sea level fluctuation curves than those generally published for the Quaternary are valid.

1. INTRODUCTION

Littoral lowlands in India were studied to find indicators for higher sea levels, date the sediments and reconstruct the geomorphological and tectonic evolution during the late Quaternary.

CHATTERJEE (1961) was one of the first to deal systematically with the effect of the Quaternary sea level fluctuations on the coasts of India. He marked the terraces on

169

D. B. Scott et al. (eds.), Late Quaternary Sea-Level Correlation and Applications, 169–194.

a 1:10,000,000 scale map and – based on altimetry – correlated them with the 'classical' Mediterranean stratigraphy concluding that "all the five well-established marine terraces and a new one have been traced on the Indian coasts", i.e. Sicilian, Milazzian, Tyrrhenian, Monastirian, late Monastirian or Cambian and Konkanian (p. 55). Since Peninsular India as part of the former Gondwanaland is supposed to be a 'stable continent', it would then be possible to find out about the altitude of the eustatically caused transgressions without any major tectonic impact.

Today, however, it is well substantiated that altimetry alone is not a useful indicator for chronosequences of marine terraces (BRÜCKNER 1980; 1985), and a worldwide transfer of the Mediterranean concept – which itself is far from being established beyond doubt (BRÜCKNER 1986) – is not to be recommended. Moreover, it turns out that most of CHATTERJEE's (1961) so-called "marine terraces" are planation surfaces of terrestrial origin. The stability is also questionable since AHMAD (1972) indicates sites which give evidence for emergence and others for submergence.

Absolute dating of the fossiliferous sites of the marine Quaternary has been lacking until now. Together with morphostratigraphy, this alone can help to answer the questions concerning eustatic sea level fluctuations and neotectonic movements. This dating has now been systematically done with the help of various dating techniques (Io/U = Th/U = ^{230}Th/^{234}U, ESR = Electron Spin Resonance, ^{14}C = Radiocarbon).

The Kathiawar Peninsula, the Konkan Coast and the coasts in Southern India were chosen for research. Since many parts of the Indian coasts are in the (semi)humid tropics where decalcification is rapid, it was not easy to find fossil material appropriate for the application of these dating methods. Most of the marine deposits presented here were dated for the first time. Derived therefrom, a reconstruction of the geomorphological and tectonic evolution of these selected areas during the late Quaternary is presented.

2. THE SAURASHTRA COAST OF KATHIAWAR PENINSULA

2.1. The Miliolite limestone

Many papers have been published on Kathiawar Peninsula[1] because of the 'Miliolite Formation' outcropping up to +60 m in a 10–20 km wide coastal belt but also in patches inland up to 247 m. This organogenic grainstone contains 90–95 % $CaCO_3$ and has a granulometric composition of 80–95 % in the sand, 10–20 (up to 40) % in the silt and 1–3 (up to 25) % in the clay fraction. The limestone owes its name to foraminifers of the family Miliolidae, a microfauna living in shallow warm water. According to the typical sites at Porbandar it is also known as 'Porbandar Stone'.

Concerning the genesis of miliolite rock there are two different interpretations: a marine and an eolian one. The dispute is documented in many articles summarized by MERH (1980) and in the papers of the NATIONAL WORKSHOP (1986). The answer to this question is the key to the conclusions concerning sea level fluctuations and neotectonics derived therefrom.

The miliolite limestones have been regarded by some authors as evidence for eustatic changes of the Quaternary sea level. Under the assumption of a marine genesis even for the inland miliolites MATHUR and VERMA (1979) postulated major marine transgressive

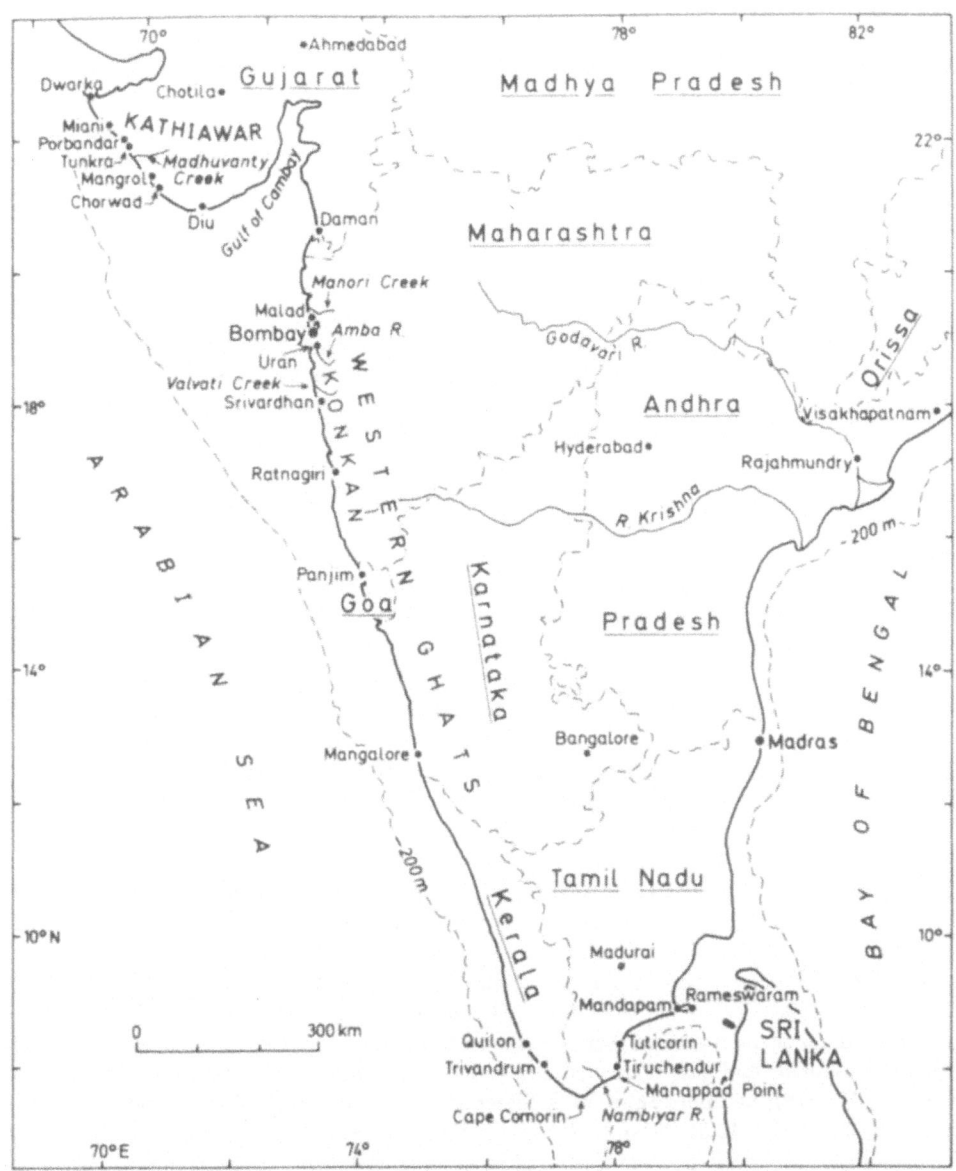

Figure 1.   Map of Peninsular India indicating the sites mentioned in the text.

cycles up to 220 m which they correlated with the 'classical' Mediterranean stratigraphy for the marine Quaternary. They took such tectonically unstable coasts as those from Chile and Morocco as references for a glacio-eustatic formation of the miliolite terraces of Kathiawar. SPERLING and GOUDIE (1975) came to a totally different result: The high-level miliolites are all of eolian origin. BASKARAN (1985) dated 48 rock specimens (by Th/U) and postulated a deposition in three episodes: around 60 ± 10 ka, 95 ± 20 ka and 170 ± 30 ka. Favouring a marine origin of most miliolites he calculated uplift rates of 2.2 m/ka in the central part of the peninsula. Even the NATIONAL WORKSHOP (1986) did not yield a unanimous result concerning the origin of these organogenic grainstones.

Based on chronostratigraphic, petrologic and paleozoological analyses BRÜCKNER et al. (1987) came to the conclusion that the inland miliolites are of eolian origin (Photo 1). They are skeletal grainstones or biosparites. The dominant type of cement is a low-magnesian blocky spar which indicates a meteoric-vadose cementation. No feature is indicative of a marine or intertidal cementation. Therefore, these miliolites are not indicators either of late Quaternary marine transgressions or of uplift.

In the coastal miliolite belt, however, marine macrofossils can be found in the basal parts of miliolite quarries. Between Porbandar and Tunkra coast parallel ridges of miliolite reach up to +15 m. In their lower parts marine macrofossils were found. The genera *Tellina*, *Cerithium*, *Ostrea* and *Arca* occurred in a quarry SE of Porbandar up to 3–4 m above sea level. Some of the *Tellina* samples were even in living position. The transition from shallow marine to a lagoonal environment is shown by the fact that the valves of *Tellina* become thinner and smaller the higher we go in the profile up to 4 m. All species discovered can adjust to a lagoonal environment. The microfaunal record reveals Foraminifera, Ostracodae and Bryozoae. They are all well preserved and do not show any features of transport.

In quarries near Tunkra a similar situation was found. Fossils (*Venus*, *Ostrea* and *Conus*), with well-rounded and flattened pebbles (mostly 2–4 cm, maximum 15 cm $\emptyset$) and a recrystallized drift coral pebble of *Favia* sp. (10 cm $\emptyset$), up to 4 m above sea level, underline the definitely marine environment. The upper parts of these beach ridges are formed by high eolianites, representing the former coastal dunes.

Dating the miliolites has been the principal problem. The Io/U and the ^{14}C dating techniques of untreated rock specimens can give mixed and therefore minimum ages only, as both the carbonates of the forams as well as the cement are dated. Therefore, they are rejuvenated. This can clearly be demonstrated: The total miliolite rock surrounding a closed *Tellina* sample at Porbandar is 61.0 ka (54.1–68.6) (sample "Por 1c-M") Th/U-yr old[1]. After the organogenic particles were separated from the cement (treated with 1:1-diluted hydrogen peroxide and passed through a 0.1 mm sieve), the result was 95.0 ka (72.6–123.2) ("Por 1c-M/1:1"), i.e. more than 50 % 'older'. The *Tellina* sample itself yielded a Th/U age of 37.3 ka (32.1–42.8) ("Por 1c-S"). From the stratigraphical point of view this is too young an age. Obviously post-depositional Uranium enrichment – quite common with very thin shells – led to this apparent younger age.

Electron Spin Resonance dates show that it was the last interglacial transgression that deposited these sediments. Shells of the basal parts of the Porbandar and Tunkra quarries are 94.9 ka ("Por 1c-S"), 105 ka ("Por 1b") and 97.6 ka ("Tun 2d") $\pm$ 10 % ESR-yr old. The miliolite of the youngest generation (Miliolite-I) was – at least in its lowermost parts – deposited during stage 5 of the oxygenisotope record. Until now macrofossils have only been found nearshore up to an altitude of 4 m above sea level. The lagoonal facies in particular proves that the maximum sea level of stage 5 was only 4 m higher than the present one. Therefore, miliolites of that generation up to +4 m may be of marine origin. Since there is no intercalated paleosol, the upper parts have to be interpreted as former coastal dunes capping the marine beach ridges.

In some profiles an older miliolite generation (Miliolite-II) underlies the younger one (Miliolite-I); they are separated by a paleosol. Within the coastal miliolite belt at least two even older miliolite generations of aquatic (?) origin are found up to 62 m above sea level (the sites are shown in BRÜCKNER et al. 1987). These miliolites are older than the upper limit of the Io/U dating technique. So far, however, except for the coastal sections at Porbandar and at Tunkra, it is only near Mangrol that macrofossils have been found in miliolite quarries. At the road from Madhavpur to Mangrol, 3 km beyond Shil, 8 km before Mangrol, the marine miliolite reaches up to 10 m above sea level. Several pieces of large *Ostrea* sp. and other shells plus a coral pebble have been found. They are all recrystallized and therefore not dateable. Compared to the well-preserved fossils up to +4 m they seem to originate from an older transgression of unknown age.

Based on geomorphologic and petrologic-criteria the base of the 10–20 km wide miliolite belt may be of marine origin. According to SHRIVASTAVA (1968, 95) miliolite beds rest unconformably on the Pliocene strata north of Porbandar. Therefore, the first marine miliolite transgression dates from the post-Pliocene. This confirms a statement by MARATHE et al. (1977; cit. by HUSSAIN et al. 1980), who postulate a marine transgression up to +40 m. As of now, the definitely marine environment, however, can only be proven with macrofossils and marine pebbles up to a maximum altitude of 10 m above sea level.

2.2. Marine terraces

Marine fossils have mainly been reported from the westernmost part of Kathiawar Peninsula. By dating molluscs and corals GUPTA (1972; 1977) concluded that at 4.5–6.7 ka, 23.0–35.1 ka and 105–125 ka the sea level was 2–6 m above the present one. While the last interglacial transgression is based on two Io/U ages of 113 ± 4 ka and 129 ± 9 ka from samples in c. +4 m, the last glacial one (23–35 ka) is based on ^{14}C dates which are not reliable on Pleistocene molluscs. The Holocene dates come from samples at altitudes of 1–6.50 m. SOMAYAJULU et al. (1985) redated the samples and could only confirm the Holocene transgression. They rejected the 30 ka high level and found the younger Pleistocene ages to scatter in the range of 118–176 ka.

Between Mangrol and Chorwad an older miliolite (Miliolite-II) forms coast parallel beach ridges. The last interglacial sea transgressed into the former depressions and accumulated marine sediments which are 86 ka (72–104) Th/U-yr old ("Mangrol 2a"). Then followed terrestrial alluvium, partly consisting of younger resorted Miliolite-I. Apparently the forams were blown into the coastal swamps and helped to silt them up (Fig. 2).

The best younger Pleistocene terrace is visible in a cliff profile at the beach of Chorwad Holiday Camp. There the miliolite is at least 200 ka (178–230) Th/U-yr old ("Chor 10a") and was built up in stage 7 or earlier. A marine terrace was cut into it with abundant macrofossils reaching up to c. 4 m above sea level and indicating stage 5 (Fig. 3, profile B).

The morphostratigraphic evidence confirms that the last interglacial sea transgressed into older beach dune ridges, partly abraded them and partly filled up the lagoonal depressions between them.

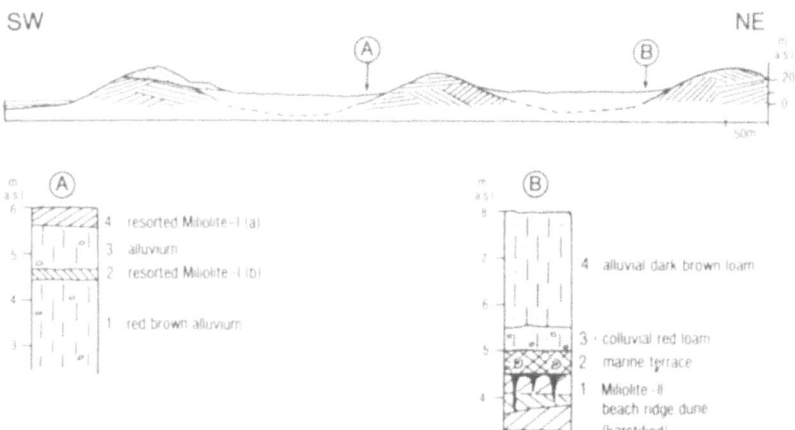

Figure 2.   Miliolite beach ridges and marine terrace at Mangrol.

In summarizing the results we can say that Miliolite-I was formed during the general regression that followed the last interglacial maximum of the Arabian Sea. Lower parts may be of marine-lagoonal origin whereas the upper parts of the beach ridges represent the former coastal dunes.

The respective error ranges of the dating techniques render a correlation of the different marine deposits with the substages 5e, 5c or 5a impossible. The last interglacial sea reached its maximum +4 m above the present MSL[2], c. 2 m above the HTL, which can be derived from the lagoonal facies of the Porbandar profiles. In this context it is worth noting that GUPTA (1977), too, found the last interglacial samples only up to this altitude. As 4 m is slightly below the assumed glacio-eustatically 'normal' position of the 125 ka-strandline (cf. BRÜCKNER 1986) no uplift seems to have taken place since then. The position of these dated fossils is a strong argument against the postulated transgression deep into Kathiawar in the late Pleistocene and the neotectonics derived therefrom.

Higher Quaternary sea levels proven by macrofossils are rare. One is found near Mangrol up to 10 m, but because of recrystallization dating is not possible. Higher and older miliolites within the coastal belt may be of marine origin up to 30–60 m; they need further research.

2.3. The Holocene transgression

As already mentioned above, a transgression around 6,000 BP is confirmed by GUPTA (1977) and SOMAYAJULU et al. (1985). The altitude of up to 6.40 m (GUPTA's sample no. TF-1059, shells from a "beachrock" near Porbandar, 6,300 ± 250 ^{14}C-yr old) is eustatically impossible and would presuppose an enormous uplift. If the sample was not contaminated, it could rather be interpreted as a storm beach.

Lothal is a famous Harappan site 80 km S of Ahmedabad, 12 m above present sea

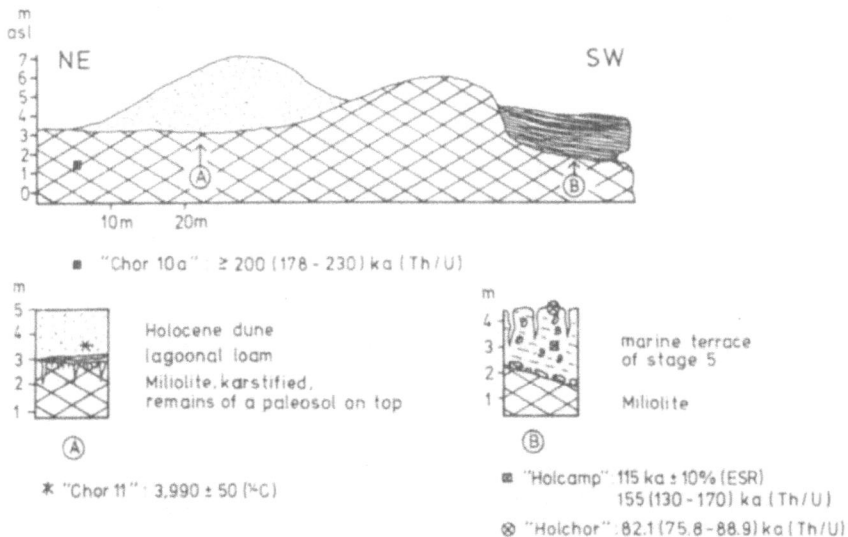

Figure 3. Geological cross-section at Chorwad Holiday Camp.

level. Between 5,500 and 3,500 BP the Harappan civilization was at its peak in this part of Gujarat. The city, now 23 km from the Gulf of Cambay, was then only 5 km away from it (S.R. RAO; cit. by PANDYA 1977). This could also be attributed to the river sedimentation, but as the gulf is extremely shallow, even a small drop of sea level causes a big retreat in the shoreline.

A profile at the mouth of Madhuvanty Creek was studied in detail (Fig. 4). Thin sections of the lowermost part ("Madu 1") show a well-sorted grainstone with a porosity of 40 %. It is rich in corals, molluscs, coralline algae, echinoderms, bryozoans, foraminifers (Amphisteginidae, Rotaliidae), micritic intraclasts and minerals from igneous rocks (quartz, plagioclases, pyroxenes, vitreous particles). The cement is a blocky mosaic of low-magnesian calcite. It seems to be a wind-blown material, probably a coastal dune, which was later diagenetically transformed into an eolianite. A red paleosol developed on it.

The marshy soil shows that around 7,600 ± 155 BP the spring tide reached 1.70 m above the present mean high tide (which is forming an upper abrasion platform in the eolianite). It represents a last stillstand of the Holocene transgression, before its highest peak around 5,000 ± 70 yr BP accumulated the marine sand (with *Cypraea* sp. among others). Loose beach sand is not very useful for properly calculating the altitude of the paleosea level. In comparison with the present situation and the position of the much more reliable former tidal flat one can conclude that the Holocene sea level was c. 1 m higher than the present one. Such a mid-Holocene sea level may well explain the existence of tidal inlets, like those at Miani Creek, that show abandoned tidal flats in their inner parts (see also BEDI and RAMANA RAO 1984).

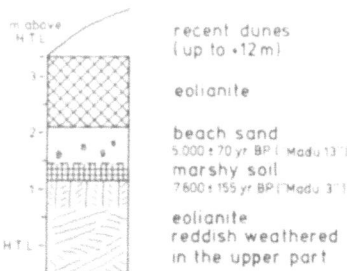

Figure 4.   Holocene deposits at the mouth of Madhuvanty Creek.

On top is the consolidated former coastal dune. "Madu 5" (Photo 2) is a well-sorted, biogenic grainstone, similar to "Madu 1" but poorer in igneous minerals. It has a porosity of 40–50 %. Initially regular rims of sparry calcite later developed into blocky mosaic cement. Laterally, micrite matrices occur at grain contact. The wind-blown material was first cemented within freshwater phreatic to vadose zones. Micrite at grain contact presumably reflects a later stage of cementation under the control of sea spray, which is still happening even today.

3.   THE KONKAN COAST

This is the coastal strip, 30–100 km wide, which extends from the Arabian Sea to the Western Ghat escarpment and from north of Bombay to north of Goa (c. 19°30'N–16°N). Field checks as well as landsat imageries (Photo 3) show that this part of the west coast is a ria-type coast where the tidal waters may penetrate up to 30–45 km inland and the river mouths are widened to broad creeks. The whole appearance is that of a submerged coast. Prominent headlands, 30–75 m high, built up by Deccan Trap basalts, are hindrances for the positive shift in the shoreline. In the small bays between them the longshore current created sand bars behind which fluvio-marine marshes with mangroves ('kharlands') evolved by the interaction of sea and river deposition. In this way these estuaries are becoming shallow. A good example is visible near Srivardhan (Fig. 6). The genesis of the littoral lowland of Konkan has just recently been described by BRÜCKNER (1987). Here we only deal with its late Quaternary evolution.

So far, no Pleistocene marine deposits have been detected along the Konkan Coast. The four "marine terraces" in the Ratnagiri District described by SUDESH KUMAR and RAO (1982) are either of fluvial origin or of Holocene age.

Lower levels of the Arabian Sea are proven by submerged terraces on the western continental shelf between 14° and 20°N in -92, -85, -75 and -55 m (NAIR 1974). It is most likely that these terraces owe their existence to stillstands in the Flandrian transgression, which followed the maximum of the last glacial regression.

According to cross-sections of the lower river courses and creeks the valley floors

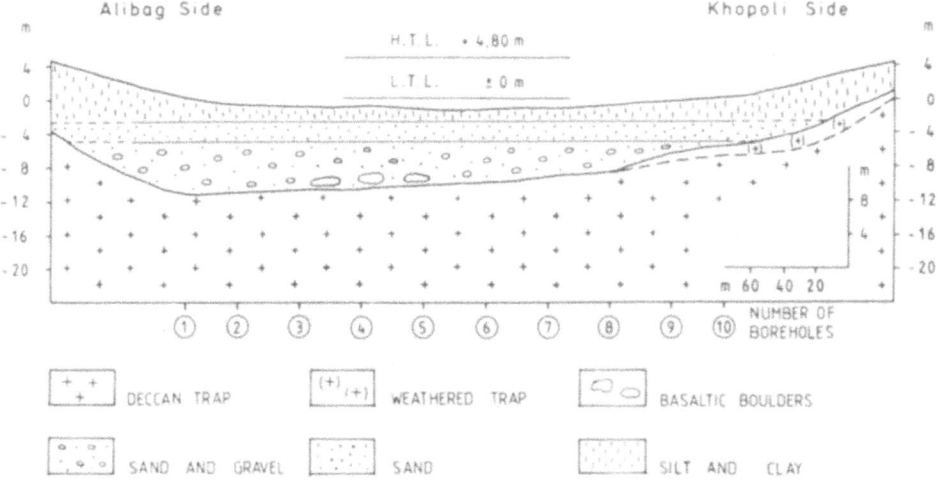

Figure 5.   Drilling profile across Amba River at Dharamtar
              Creek bridge of S.H. 42 (superelevation: 5 x).
              (source: P.W.D. Bombay, Subdivision Alibag, 1958)

are relatively flat (BRÜCKNER 1987, Figs. 4a–4c). Their lowest positions of -2 m (Karanja Creek), -8 m (Revadanda Creek) and -11 m (Amba River) show an incline to a deeper sea level. Moreover, the incline results from submergence, leading to this ria-type coast. The fact that not even last interglacial marine deposits are found onshore is also a sign of subsidence.

The Amba profile at the bridge crossing Dharamtar Creek is especially interesting (73°01'50"E, 18°41'49"N, SW-NE profile; Fig. 5). The coarse sediments at the base indicate a higher fluvial energy than today, probably deposited during the last glacial period when the level of the Arabian Sea was much lower. From the sand layer in c. -3 m below LTL a dredger brought up marine fossils (large oyster shells, snails, molluscs, coral pieces) which were ^{14}C-dated: sample "Amba 2" is 2,600 ± 60 yr old. During that time a marine transgression took place so the sand layer can be attributed mostly to a marine environment. The upper silts and clays are river marsh deposits.

This interpretation is confirmed by GUZDER (1980), who researched many fossiliferous sandstone occurrences (locally called 'karal') along Konkan. Based on radiocarbon dates she concluded that around 2,500 BP the sea level was c. 1 m higher than today. This littoral 'concrete' occurs in broken stretches from Daman (20°24'50"N) to Goa (15°36'N). Several sites were researched in detail.

North of the mouth of Manori Creek (19°12'10"N) a sandspit was formed between two hills during the Holocene transgression. It extends over 3 km, and its maximum height is 3–5 m above HTL. AGRAWAL and GUZDER (1972, 219) obtained ^{14}C ages of 4,245 ± 85 BP (in -1.0 m), 4,540 ± 100 BP (+1.0 m) and 4,385 ± 110 BP (+3.0 m).

A thin section of a sample from the upper part of the beach ridge shows a moderately sorted grainstone with high porosity of 50–60 % where molluscan detritus is the main sediment contributor. Subordinate grains include Amphisteginidae, encrusting foraminifers, alcyonarian spicules, diagenetically formed carbonate grains, minerals and fragments from igneous rocks (= Deccan Trap). The cement shows a blocky mosaic and 'whisker' crystals (Photo 4). Therefore, this coastal deposit (beach or coastal dune) was secondarily cemented within a freshwater vadose zone. It is not an intratidal formation.

A similar friable shell limestone outcrops at the mouth of Khatan (= Valvati) Creek south of Arawi, where it forms an ancient beach ridge up to 2 m above HTL and 100–150 m farther inland than the present one (Fig. 6, Photo 5). Prospection pits show a thickness of more than 4 m, and it can also be traced in well sections of Arawi.

The macrofauna *Arca*, *Aentalium* (at least two different species) and *Donax* could be determined. Two shell samples were taken from the central prospection pits: "Valvati 1" (1 m above HTL and 1 m below surface) is 2,260 ± 60 BP and "Valvati 3" (2 m above HTL, surface sample) is 2,410 ± 60 BP ^{14}C-yr old. Since then, a regressive shift in the shoreline of 100–150 m has taken place.

NW of Ratnagiri the karal forms a fossil beach ridge, connecting the island Mirya with the mainland. Its uppermost parts reach up to 4 m above HTL. They are karstified (with 1.50 m long karst pipes) and covered by a brown soil, 10 cm thick. Above this are a loose fine dune sand (20 cm) and an anthropogene layer (20 cm, with a lot of potsherds).

The western side of this beach ridge is exposed to the open sea. Here in karal pits small specimens of *Donax* and *Venus* were found. In 3.50 m above HTL, they were dated: 2,030 ± 50 BP ("Mirya-F"). AGRAWAL and GUZDER (1972) published ^{14}C ages of 2,800 ± 110 BP and 2,305 ± 95 BP in +5.9 m and +6.0 m (above MSL). Being an eolianite, the paleosea level cannot be derived from this upper part. It is similar to the one at Madhuvanty Creek (chapter 2.3). Thin sections show a moderately sorted grainstone with 25 % porosity. Half of the material consists of skeletal, the other half of igneous particles. The skeletal fraction includes Miliolidae, Amphisteginidae, Mollusca, coralline Algae, encrusting Foraminifera, Bryozoae and diagenetically formed carbonate grains. Quartz and vitreous particles dominate in the igneous fraction. They are encrusted by a blocky mosaic cement. Only the lowermost parts of this up to 5 m thick formation were formed in an intertidal environment, the upper parts show dune features.

On the west coast of the Uran Peninsula (SE of Bombay) the present abrasion platform is developed in the Deccan Traps. A fossil beach, 3.8 km NW of Uran, starts with beach pebbles and many remains of fossils (corals, oysters etc.), forming a beachrock up to 0.50 m above HTL. The strata are covered by anthropogenetically disturbed sands, 1 m thick. From the fossil beach we obtained a radiocarbon age of 1,470 ± 55 BP (sample "Uran-D").

On the same coast, 1.5 km south, the profile shown in Fig. 7 is visible in a long cliff section. *Cerithium*, *Oliva*, *Dentalium*, *Arca*, *Trochus*, *Nassa*, *Cyclope*, many coral fragments and numerous specimens of *Nerita* are present. The latter may live in a shallow sea and can adjust to brackish conditions. *Nerita* from the uppermost part of this fossil beach is 1,800 ± 50 ^{14}C-yr old ("Ur 10"). An ESR dating yielded a young Holocene

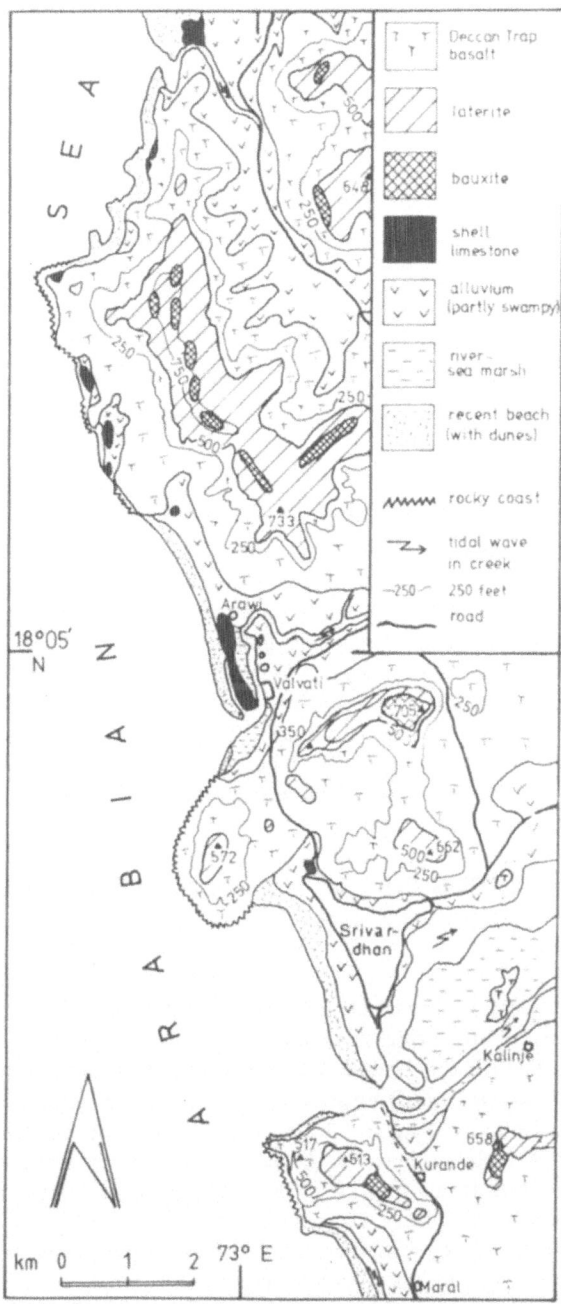

Figure 6.   Geological map of the area around Srivardhan showing the occurrences
of shell limestone (= 'karal').

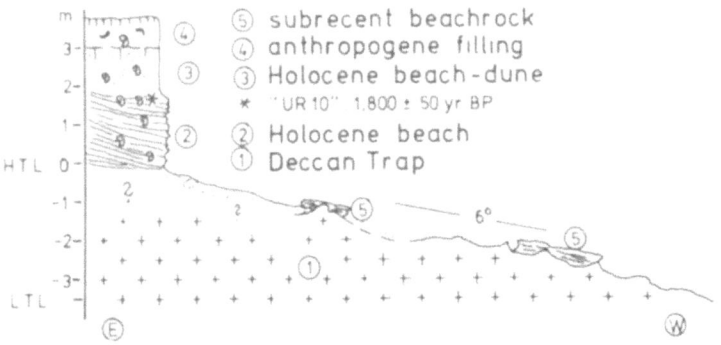

Figure 7.   Holocene beach at Uran.

spectrum.

Holocene beach sediments up to 2 m above HTL are extraordinary. Yet we have to consider that presently at Bombay the average tidal range is 3.4–4.6 m, the mean range of spring tides even 5.7 m. We do not know about the Holocene ones. The data show how difficult it is to reconstruct the position of the paleosea level from this fossil beach at Uran.

The existence and the dates of the karal formation prove that the Holocene transgression reached a maximum c. 2,000–2,800 BP and left fossil beach ridges along the Konkan coast. The comparatively too young ages of the Uran samples may be caused by contamination with more recent HCO_3^-, as unlike the other samples they are reached by seawater spray. The limited data base prevents an overinterpretation in the sense of an even younger transgression or differential tectonics.

The altitude of the Holocene transgression is a problem. Different articles in *Sea Level Research* (ed. by VAN DE PLASSCHE 1986) point out how difficult it is to find appropriate indicators for an exact reconstruction of the paleosea level. In the case of beach ridges this is nearly impossible, all the more as the upper parts consist of dune facies. The position of the beach ridge at Valvati (100–150 m behind the present one) may be a hint for a regression since its formation. From the position of the lagoonal *Nerita* specimens at Uran (c. 0.80 m above the present spring tide) and the analogy to southern India we conclude that around 2,000–2,800 BP the Arabian Sea was at the most 1 m higher than at present.

In the western part of Uran Peninsula the relict of a young storm beach is preserved. On an older abrasion platform, up to 2.30 m above HTL, a 30 cm thick marine deposit consists of loose, unsorted beach material (basalt pebbles, debris, fossils) plus charcoal and potsherds in a dark brown loamy sand. It is 180 ± 50 ^{14}C-yr old ("Uran-S") and was later covered by a 30 cm thick debris layer. Based on its chaotic stratigraphy and exposure to the Arabian Sea it can be interpreted as the product of a stormflood occurring when tropical cyclones hit the coast. Such a 'hundred-year flood' can strongly modify the littoral zone both in an abrasive and an accumulative way. It may well reach higher altitudes than the eustatically proven Holocene fluctuations.

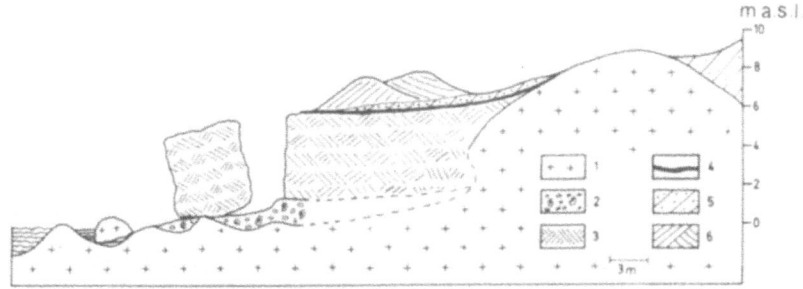

Figure 8.   Profile of the last interglacial marine terrace at Cape Comorin.
1 = Precambrian bedrock, 2 = marine terrace, 3 = eolianite,
4 = calcrete, 5 = teri sands, 6 = recent dunes

4.   THE COASTAL EVOLUTION OF SOUTHERN INDIA DURING THE LATE QUATERNARY

4.1. From Cape Comorin to Rameswaram

4.1.1. The Pleistocene Epoch

Research was carried out on the coasts between Trivandrum and Rameswaram. As at Konkan, no marine deposits of the early or middle Quaternary were found, only sites of the 125 ka-strandline and of the Holocene.

At Cape Comorin, the southernmost part of Peninsular India, the sea transgressed into an inselberg landscape: 6 km west of the cape the strata shown in Fig. 8 outcrop in a cliff section. On top of the crystalline basement a marine conglomerate is found with many shells (partly closed) and snails of the genera *Cardium* (*Trachycardium*, *Vasticardium*), *Trochus*, *Cypraea*, *Conus*, *Oliva* and *Isognomon* and big pebbles at the base. During low tide even the bottom parts are visible, but only the upper 30 cm are above HTL. Being 112 ka (98-128) Th/U- and more than 46,600 (96 % probability) ^{14}C-yr old ("Cape C.") it is a last interglacial beach. Without an intercalated paleosol an eolianite lies directly on it. This former coastal dune fossilized the marine terrace immediately after the sea had left this site.

Then follows a calcrete with a lot of encrusted snails (*Ariophantidae* cf. *Macrochlamys*) indicating a definite terrestrial environment. They are partly covered by red sands, locally called 'teri' sands (= 'sand wastes' in Tamil). The teri formation was TL-dated in Sri Lanka by SINGHVI et al. (1986) to be of the last glacial epoch.

The terminal point of the transgression does not outcrop, but there is no doubt that it is less than 2 m above HTL. Definitely no uplift but rather a slight submergence of the coast has occurred since 125 ka.

Where the Nambiyar River flows into the Gulf of Mannar, 35 km ENE of Cape Comorin, a former beach with abundant macrofossils is being partly eroded by the river. On top of intensely weathered gneiss we find marls, indicative of a marshy environment. Then follows a calcite cemented marine sand, up to 2.50–3 m above HTL. The marine terrace is 124 ka (112–138) Th/U-yr old ("Vija 52"), a calcrete on its top 87.5 ka ± 10% ESR-yr old or younger ("Vija 53")[3]. The terrace was later karstified and the karst depressions then filled with teri sands.

After the formation of the marine terrace in stage 5, a semiarid climate created the limestone crust in the period between the last interglacial and the last glacial epochs or later. Then the climate became wetter so that the crust was partly dissolved. This process ceased at least in the glacial maximum when wind-blown teri sands fossilized the landscape.

A similar stratigraphy as the one at Cape Comorin is found still farther north, at Manappad Point, 16 km SW of Tiruchendur: A fossil beach, up to 3 m above HTL, grades continuously into a 30 m thick eolianite which is topped by a calcrete. Again this beach was deposited by the 125 ka-strandline: An in situ sample of *Balanus* sp., 1.25 m above HTL and 120 ka (110–127) Th/U-yr old ("Tuti 2b"), is a good bioindicator for the definitely higher paleosea level.

Finally, another last interglacial site is observed on the island of Rameswaram at Adam's Bridge connecting India with Sri Lanka. Its northern part is a fossil coral reef (Photo 6; see also STODDART and GOPINADHA PILLAI 1972). A coral sample of *Porites* sp. was taken from the middle of the island, 15 cm below surface, c. 2.50 m above LTL: 112 ka (107–120) (Th/U), 139 ka ± 10% (ESR) ("Vadakadu 4"). As the low tide is the upper level for present-day corals to grow, the present position shows that this last interglacial reef underwent a slight subsidence rather than an uplift.

In the southern part of Peninsular India the main transgression of the last interglacial sea left patches of fossil beaches only on its eastern side from Cape Comorin to Rameswaram.

Normally it is not possible to differentiate between the substages of stage 5 on the basis of 'absolute' dates alone (BRÜCKNER 1986). In our case, all dates are from the earlier part of stage 5, which suggests a formation in stage 5e. If they were of the substages 5c or 5a, then 5e should be found in a higher position, which is not the case. The geomorphological record shows that this last interglacial terrace is the only one, and therefore has to be attributed to the 125 ka-strandline.

It is difficult to derive a paleosea level from marine deposits unless bioindicators are found. The corals of Rameswaram and the *Balanus* sample at Manappad Point are suitable for that purpose. The other sites may have been deposited during spring tides or even storms. That they are all found only up to c. 3 m above the present HTL (the tidal range in the south is around 1 m) signifies a subsidence, as the eustatically 'normal' position of stage 5e is assumed at c. 6 m above the present mean sea level (see e.g. STEARNS 1976, BRÜCKNER 1986).

4.1.2. The Holocene Epoch

The area around Mandapam is the only region in the south where a Holocene transgression deep inland can be proven.

This extensive peninsula is built up by marine facies. Where it is exposed to the

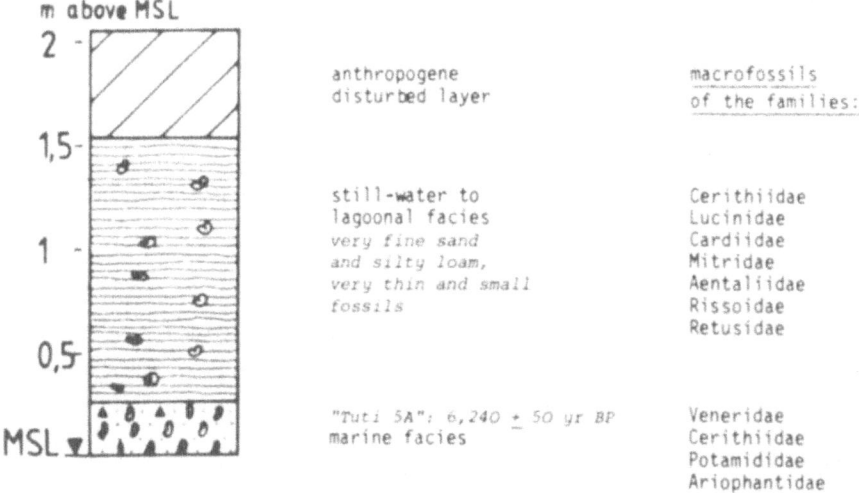

Figure 9.   Holocene beach and lagoonal deposits north of Manappad Point.

SW-monsoon, up to 19 m high dunes have developed. In its center grey marine sands and lagoonal loams occur up to 1 m above HTL, indicating a transition from shallow marine to lagoonal environments. The sand contains abundant closed bivalves (Photo 7). About 4 km W of Mandapàm a sample of *Cardium* sp. in living position is 2,740 ± 60 BP ^{14}C-yr old ("Man-M").

East of Mandapam, the northern side of this peninsula is formed by a beachrock consisting of pebbles (up to 15 cm Ø) and many fossils, up to 0.50 m above HTL. A drift coral fragment in it yielded a ^{14}C age of 3,660 ± 65 yr BP ("Man-C"). During monsoon time the sea may reach the upper parts of the beachrock so its position need not imply a higher sea level during the time of deposition.

N of Manappad Point, between Kulasekarapatnam and Tiruchendur, a fossil beach ridge, up to 7 m above sea level, c. 1 km inland, separates a former lagoon from the sea. Later the lagoon dried up and became landlocked. A profile (Fig. 9) shows a rich marine fauna up to 20 cm above MSL, deposited in a shallow marine environment. The layer is closed up by a 5 cm thick sand with a lot of garnet minerals and a few terrestrial snails (*Ariophantidae* cf. *Macrochlamys*), proving a beach environment. This beach is 6,240 ± 50 ^{14}C-yr old ("Tuti 5A"). The following very finely laminated silt and fine sand layer is a calmwater facies and was deposited during the main Holocene transgression.

What was an open bay and later a beach in the early middle Holocene turned into a lagoon when the barrier spit closed the bay. That these lagoonal deposits on top of the former beach reach up to 1 m above HTL indicates a transgression after c. 6,200 BP.

4.2. The southern coast of Kerala

The west coast of southern India shows definite signs of subsidence. It is the picturesque

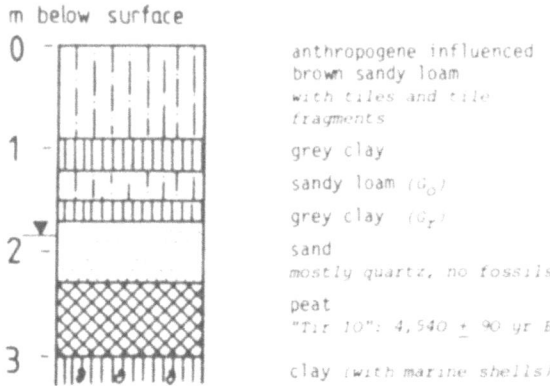

Figure 10.   Peat layer in the brickworks at Tiruvallam near Trivandrum.
Arrow indicates MSL.

backwaters of Kerala where rivers are delayed on their way to the sea because of an active
longshore current building up beach ridges. This part of the Malabar coast consists of
parallel beach ridges separated by lagoonal areas and wide river courses.

Until now no Pleistocene shorelines have been found. Even the highest Holocene
marine deposits are drowned.

Fig. 10 shows a profile in the brickworks near Tiruvallam at Trivandrum. The groundwater level corresponds to the MSL. Marine clays are found up to -115 cm. On their top
a peat developed 4,540 ± 90 BP ("Tir 10") up to -45 cm. It was later covered by fluvial
sands and clays which are devoid of fossils.

The middle Holocene transgression deposited the lagoonal clay. Then, in a marshy
environment, the peat was able to grow.

At Kadiapattanam, 30 km WNW of Cape Comorin, a new bridge crossing Valliyar
River was built in 1986. Being less than 2 km from the coast the river is influenced by
the tidal effect. Trial drillings for the bridge pillars revealed a 1.2 m thick layer of fossil
wood and plants (9.20 to 8 m below MSL) on top of lagoonal-marine (?) clay and sand
layers. The fossil wood is 7,130 ± 65 ^{14}C-yr old ("Kadia-HK"). The upper 8 m consist of
deposits devoid of fossils: 7 m of loam with 1 m of sand on top. The profile indicates a
subsidence of the west coast at least in the late Quaternary Epoch and a subsequent fill of
river alluvium.

5.   CONCLUSIONS

Littoral lowlands in India were studied to find indicators for higher sea levels, date the
sediments and reconstruct the geomorphologic and tectonic evolution during the late
Quaternary.

On the Kathiawar Peninsula, the miliolite problem (marine versus eolian genesis) could be solved by micropaleozoological, thin-section and ^{230}Th/^{234}U analyses. In near-shore profiles the lower parts (up to 4 m above sea level) of these organogenic grainstones were deposited by the 125 ka sea level, whereas the upper strata and the interior milio-lites were wind-transported after the maxima of the penultimate and the last interglacial transgressions, respectively.

Along the Konkan Coast, all coastal processes took place on the early Tertiary Trap basalts. Shorelines of the Pleistocene are missing. This and the ria-type coast indicate subsidence. As for the Holocene, ^{14}C-dated beach sand barriers prove a higher sea level around 2,500 BP.

In southern India deposits from the last interglacial transgression reaching up to +4 m were for the first time dated by ^{230}Th/^{234}U and ESR. They are all from stage 5e. This part of the former Gondwanaland may have subsided slightly since 125,000 BP. The Holocene transgression reached up to 1 m (^{14}C-dated sandbars and lagoons).

Deposits from the early and middle Quaternary are missing. It seems that the published curves of the Quaternary sea level fluctuations which show a general decrease of the maxima of the transgressions during the interglacial epochs are only valid for geologically unstable uplift regions like the Mediterranean Basin. Interestingly enough, it was there that the stratigraphy for the Quaternary marine terraces was originally established (see BRÜCKNER 1985). For geologically old continents, like Peninsular India, different curves seem to be needed.

ACKNOWLEDGEMENTS

The absolute datings were carried out by Dr. H. Erlenkeuser, Kiel (^{14}C), Dr. R. Hausmann, Cologne (Th/U), Dr. B. Kromer, Heidelberg (^{14}C), Dr. A. Mangini, Heidelberg (Th/U), Dr. U. Radtke, Düsseldorf (ESR). Dr. L. Montaggioni, Ste Clotilde, France-D.O.M., interpreted the thin sections. Prof. Dr. R.W. Fairbridge, New York, and Prof. Dr. P.A. Pirazzoli, Paris, reviewed the manuscript and made useful comments. I thank them all for their help. Two field trips were financially supported by "Deutsche Forschungsgemeinschaft" (German Research Fund), which I gratefully acknowledge.

FOOTNOTES

[1] All geographical names mentioned are shown in Fig. 1, all absolute dates in Tables I–III.
[2] MSL = mean sea level, HTL = mean high tide level, LTL = mean low tide level.
[3] Why ESR ages of calcretes have to be interpreted as maximum ages only is discussed in RADTKE and BRÜCKNER (1988).

REFERENCES

AGRAWAL, D.P., and GUZDER, S.J. (1972): 'Quaternary Studies on the Western Coast of

India: Preliminary Observations.'- *The Palaeobotanist*, **21**(2), 216–222.

AHMAD, E. (1972): *Coastal Geomorphology of India*.- 222 p., New Delhi.

BASKARAN, M. (1985): *Radiometric, Mineralogical and Trace Elemental Studies on the Saurashtra Quaternary Carbonate Deposits: Implications to their Age and Origin*.- Thesis at the Physical Research Laboratory Ahmedabad, 173 p., Ahmedabad.

BEDI, N., and RAMANA RAO, K.L.V. (1984): 'Nature and Evolution of a Part of Saurashtra Coast, Gujarat, India.'- *Z. Geomorph. N.F.*, **28**(1), 53–69, Berlin-Stuttgart.

BRÜCKNER, H. (1980): 'Marine Terrassen in Süditalien. Eine quartärmorphologische Studie über das Küstentiefland von Metapont.' – *Düsseldorfer Geographische Schriften*, **14**, 235 p., Düsseldorf.

BRÜCKNER, H. (1985): 'New Aspects of the Coastal Geomorphology.'- *Transactions of the Institute of Indian Geographers*, **7**(1), 9–16, University of Poona, Pune.

BRÜCKNER, H. (1986): 'Stratigraphy, Evolution and Age of Quaternary Marine Terraces in Morocco and Spain.'- *Z. Geomorph. N.F., Suppl.-Bd.*, **62**, 83-101, Berlin-Stuttgart.

BRÜCKNER, H. (1987): 'New Data on the Evolution of Konkan (Western India).'- In: DATYE, V.S., DIDDEE, J., JOG, S.R., and PATIL, Ch. (eds.)(1987): *Explorations in the Tropics*. (Prof. K.R. Dikshit Felicitation Volume).- 333 p., Pune; here: pp. 173-184.

BRÜCKNER, H., MANGINI, A., MONTAGGIONI, L., and RESCHER, K. (1987): 'Miliolite Occurrences on Kathiawar Peninsula (Gujarat) - New Results from Chronostratigraphical, Petrological and Paleozoological Analyses.'- *Berliner geographische Studien*, **25**, 343-361, Berlin.

CHATTERJEE, S.P. (1961): 'Fluctuations of Sea Level around the Coasts of India during the Quaternary Period.'- *Z. Geomorph. N.F., Suppl.-Bd.*, **3**, 48-56, Berlin-Stuttgart.

GUPTA, S.K. (1972): 'Chronology of the Raised Beaches and Inland Coral Reefs of the Saurashtra Coast.'- *Journal of Geology*, **80**(3), 357-361, Chicago.

GUPTA, S.K. (1977): 'Quaternary Sea-Level Changes on the Saurashtra Coast.'- *Ecology and Archaeology of Western India, Concept*, 181-193, Delhi.

GUZDER, S. (1980): *Quaternary Environment and Stone Age Cultures of the Konkan, Coastal Maharashtra, India*.- 101 p., Pune.

HUSSAIN, N., BHANDARI, N., RAMANATHAN, K.R., and SOMAYAJULU, B. (1980): 'Age of Saurashtra Miliolites by U-Th Decay Series Methods: Possible Implications to their Origin.'- *Proceedings of the Indian Academy of Sciences (Earth and Planetary Sciences)*, **89**(1), 23-29.

MATHUR, U.B., and VERMA, K.K. (1979): 'On the Quaternary Geology of Southern Saurashtra Coast Bordering Diu Island.'- *Geological Survey of India, Misc. Publ.*, **45**, 255-262, Hyderabad.

MERH, S.S. (1980): 'The Miliolite Problem.'- *Proceedings of the Sixtyseventh Session of the Indian Science Congress Association. Section of Geology and Geography* (Presidential Address), 15-42, Calcutta.

NAIR, R.R. (1974): 'Holocene Sea-Levels on the Western Continental Shelf of India.'- *Ind. Acad. Sciences*, Proceedings, B, **79**(5), 197-203. Bangalore.

NATIONAL WORKSHOP *on Quaternary Carbonates and Miliolite Problems of Gujarat, February 4-6, 1986. Expanded Abstracts*.- Physical Research Laboratory Ahmedabad (1986), 71 p., Ahmedabad.

PANDYA, S. (1977): 'Lothal Dockyard Hypothesis and Sea-Level Changes.'- In: AGRAWAL, D.P., and PANDE, B.M. (eds.): *Ecology and Archaeology of Western India*, 1977, 99–104, Delhi.

RADTKE, U., and BRÜCKNER, H. (1988): 'Problems Encountered with Absolute Dating (U-Series, ESR) of Spanish Calcretes.'- *Quaternary Science Reviews*, Oxford-New York-Beijing (in press).

SHRIVASTAVA, P.K. (1968): 'Petrography and Origin of Miliolite Limestone of the Western Saurashtra Coast.'- *Journal Geol. Soc. India*, 9(1), 88–96, Bangalore.

SINGHVI, A.K., DERANIYAGALA, S.U., and SENGUPTA, D. (1986): 'Thermoluminescence Dating of Quaternary Red-Sand Beds - A Case Study of Coastal Dunes in Sri Lanka.'- *Earth and Planetary Science Letters*, 80, 139–144, Amsterdam.

SOMAYAJULU, B.L.K., BROECKER, W.S., and GODDARD, J. (1985): 'Dating Indian Corals by U-Decay-Series Methods.'- *Quat. Res.*, 24, 235–239, New York-London.

SPERLING, C.H.B., and GOUDIE, A.S. (1975): 'The Miliolite of Western India - A Discussion of the Aeolian and Marine Hypotheses.'- *Sedimentary Geology*, 13, 71–75, Amsterdam.

STODDART, D.R., and GOPINADHA PILLAI, C.S. (1972): 'Raised Reefs of Ramanathapuram, South India.'- *The Institute of British Geographers, Transactions*, 56, 111–125, London.

STEARNS, C.E. (1976): 'Estimates of the Position of Sea Level between 140,000 and 75,000 Years ago.'- *Quat. Res.*, 6, 445–449, New York-London.

SUDESH KUMAR, S., and RAO, G.V. (1982): 'Geomorphology and Quaternary History of the West Coast, Ratnagiri District, Maharashtra.'- In: MERH, S.S. (ed.)(1982): *First National Seminar on Quaternary Environments (Baroda, November 1977), Papers*; 311–318, Delhi.

VAN DE PLASSCHE, O. (ed.)(1986): *Sea-Level Research: A Manual for the Collection and Evaluation of Data*.- Geo Books, 618 p., Norwich-Great Yarmouth.

sample	U-238 (ppm)	^{234}U/^{238}U	Th-232 (ppm)	^{230}Th/^{234}U	^{230}Th/^{232}Th	Io/U-age (ka)	Io/U-age (ka) (corrected)	comment
Holcamp (HD-1046)	1.30 ± 0.04	1.16 ± 0.04	0.01 ± 0.008	0.78 ± 0.05	—	155 (130-170)	—	shell
Holchor (KU-707)	2.438 ± 0.076	1.17 ± 0.033	—	0.547 ± 0.028	32.041 ± 7.35	83.9 (77.4-90.9)	82.1 (75.8-88.9)	shell
Chor 10a (HD-1802)	0.86 ± 0.02	1.11 ± 0.04	0.34 ± 0.02	0.96 ± 0.04	—	292 (246-∞)	200 (178-230)	miiiolite
Mangrol 2a (HD-1059)	0.364 ± 0.02	1.27 ± 0.09	0.313 ± 0.02	0.77 ± 0.06	—	148 (126-172)	86 (72-104)	Ostrea sp.
Por 1c-S (KU-703-1)	1.978 ± 0.112	1.212 ± 0.08	—	0.302 ± 0.035	24.626 ± 24.745	38.6 (33.3-44.2)	37.3 (32.1-42.8)	Tellina sp.
Por 1c-M (KU-705)	1.732 ± 0.065	1.085 ± 0.034	—	0.438 ± 0.038	47.198 ± 32.748	62.0 (55.0-69.6)	61.0 (54.1-68.6)	miiiolite
Por 1c-M/1:1 (KU-703-2)	2.162 ± 0.057	1.034 ± 0.026	—	0.615 ± 0.094	8.675 ± 4.055	102.9 (79.3-133.7)	95.0 (72.6-123.2)	miiiolite (modified)
Cape C. (HD-1064)	1.05 ± 0.03	1.12 ± 0.03	0.07 ± 0.01	0.65 ± 0.05	—	112 (98-128)	—	shell
Vija 52 (HD-1061)	0.153 ± 0.006	1.08 ± 0.04	0.014 ± 0.003	0.688 ± 0.04	—	124 (112-138)	—	shell
Tuti 2b (HD-1789)	0.65 ± 0.014	1.22 ± 0.04	0.01 ± 0.002	0.68 ± 0.03	—	120 (110-127)	—	Balanus sp.
Vedakadu 4 (HD-1787)	2.86 ± 0.04	1.09 ± 0.02	0.062 ± 0.01	0.65 ± 0.02	—	112 (107-120)	—	Porites sp.

Table I.   Ionium/Uranium Dates.
corection of the "HD" samples according to the formula: $- \left(^{232}\text{Th} \times \frac{3\,\text{ppm U}}{12\,\text{ppm Th}} \times 0.738 \right)$

sample	AD (Gy)	int. dose (mGy/a) (α, β^{-})	ext. dose (mGy/a) $(\beta^{-}, \gamma,$ cosmic ray)	k-value	thickness of sample (mm)	removed thickness (mm)	ESR-age (ka) ± 10 %	comment
Holcamp (D-I-4)	81.6	0.3157	0.3942	0.1	4.0	0.4	115	shell
Por 1b (D-1277)	65.7	0.4133	0.213	0.1	1.1	0.2	105	shell
Por 1c-S (D-1278)	75.6	0.4187	0.3793	0.1	1.8	0.3	94.9	*Tellina* sp.
Tun 2d (D-1292)	79.9	0.4027	0.4163	0.1	1.1	0.3	97.6	shell
Vija 53 (D-I-21)	57.5	0.3598	0.30	0.15	—	—	≤ 87.5	calcrete
Vadakadu 4 (D-I-30)	137.2	0.5864	0.40	0.06	—	—	139	*Porites* sp.

Table II.   Electron Spin Resonance Dates.

sample	$\delta\,^{13}$C (‰)	^{14}C (%)	^{14}C (yr BP)	comment
Amba 2 (HD 10846-10663)	−9.46	68.7 ± 0.5	2600 ± 60	shell[+]
Valvati 1 (HD 10843-10645)	0.05	71.7 ± 0.5	2260 ± 60	shell[+],
Valvati 3 (HD 10844-10698)	−0.44	70.4 ± 0.6	2410 ± 60	shell[+]
Mirya-F (HD 10978-10713)	0.09	73.8 ± 0.5	2030 ± 50	shell[+]
Uran-D (HD 9502-9323)	0.04	79.1 ± 0.5	1470 ± 55	shell[+]
Uran-S (KI-2340.001)	−2.92	97.8 ± 0.6	180 ± 50	shell
Cape C. (HD 8947-8685)	—	0.1 ± 0.1	> 50000 (1 σ) > 46600 (2 σ)	shell
Man-M (KI-2338.001)	0.03	71.1 ± 0.5	2740 ± 60	shell
Man-C (KI-2339.001)	−0.97	63.4 ± 0.5	3660 ± 65	coral
Tuti 5A (HD 10981-10722)	−1.31	43.7 ± 0.3	6240 ± 50	shell[+]
Kadia-HK (HD 10979-10732)	−29.01	41.2 ± 0.3	7130 ± 65	wood
Tir 10 (HD 10982-10733)	−27.87	56.8 ± 0.63	4540 ± 90	peat
Chor 11 (HD 10984-10735)	0.81	57.8 ± 0.4	3990 ± 50	marine snail[+]
Ur 10 (HD 10845-10646)	0.60	75.9 ± 0.5	1800 ± 50	marine snail[+]
Madu 3 (HD 10497-10350)	−9.65	38.8 ± 0.8	7600 ± 155	paleosol
Madu 13 (HD 10985-10739)	1.35	51.0 ± 0.4	5000 ± 70	shell[+]

Table III.   Radiocarbon Dates.   ([+] equivalent to 95 % of the dating standard)

Photo 1.   Quarry in miliolite dune near Una.

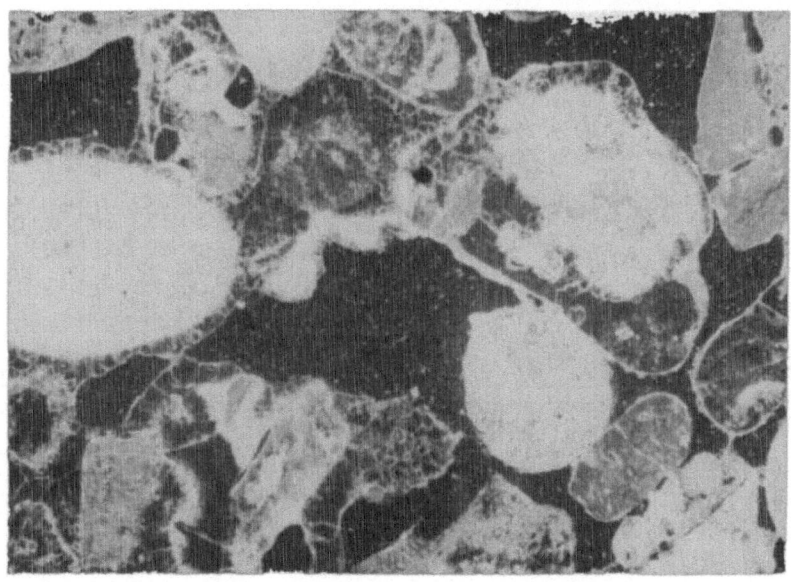

Photo 2.   Thin section of sample "Madu 5" (upper eolianite)
showing granular spar surrounding skeletal debris.
Large residual pores are visible. (120 x)

192

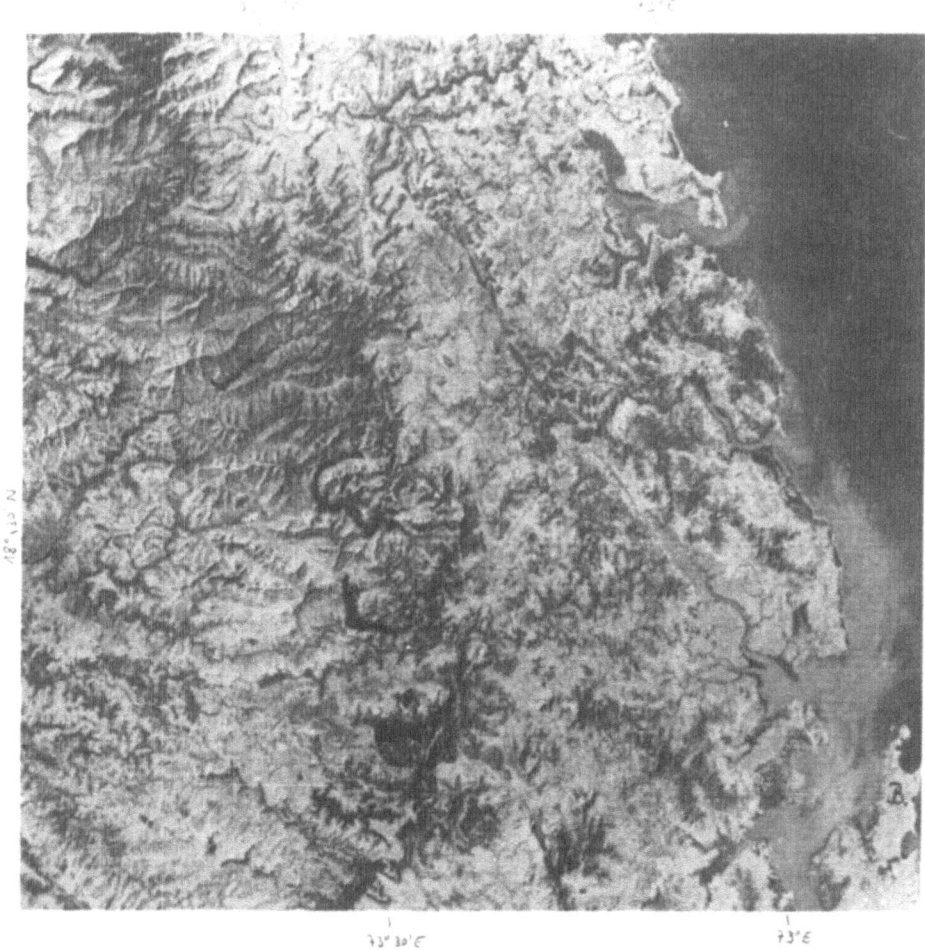

Photo 3. Landsat imagery (19 Feb. 1977) of the Konkan region south of Bombay.
The Western Ghat escarpment being the waterdivide and the wide river
estuaries are clearly visible (B. = Bombay).
(For reasons of plasticity, south is at the head of the photograph.)

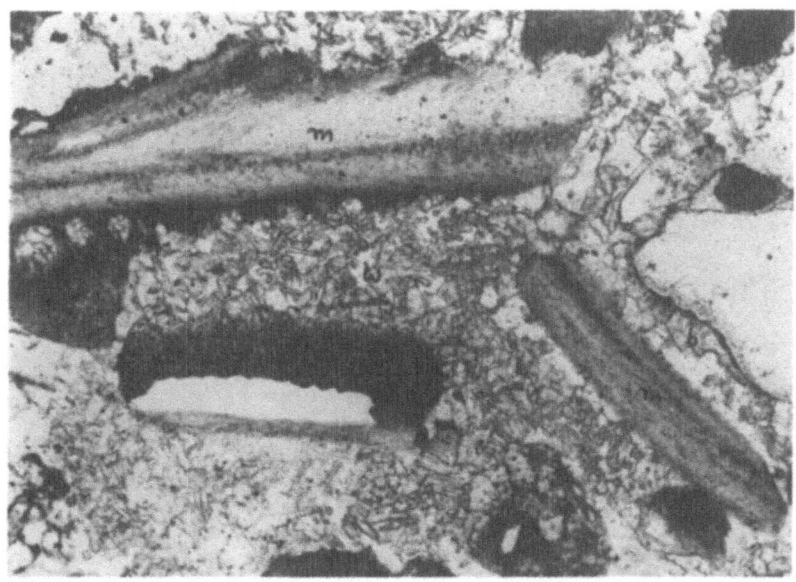

Photo 4.   Thin section of the upper part of the Manori beach ridge. (120 x)
m = mollusc, b = blocky mosaic cement, w = whisker crystals

Photo 5.   Distal part of the Valvati beach ridge, being eroded by the Khatan Creek.

Photo 6.   Fossil coral reef of Rameswaram, dunes in the background.

Photo 7.   Holocene marine-lagoonal deposits around Mandapam.

ARCHAEOLOGY AND SEA-LEVEL CHANGE IN THE SOUTHWESTERN PACIFIC: NO SIMPLE
STORY

A J Smith, Foundation Professor of Geology
Royal Holloway & Bedford New College
(University of London)
Egham Hill, Surrey TW20 0EX
Great Britain.

ABSTRACT. Late Quaternary low stands of sea level, and particularly
those which preceded the Holocene transgression, are believed to have
major significance in influencing the migration of indigenous peoples.
This is thought to be particularly true in the southwestern Pacific
where chains of islands offer, at first impression, convenient
'stepping stones'. However, the evidence currently available casts
doubts on the ability of contemporary would-be colonists to cross the
gaps which remained even when sea level was at its lowest.
    Migrations did take place, thus we must ask if the evidence is
complete. Do we wrongly start from the remains of the dispossessed?
Is the evidence only a degraded descent from the best? Much of the
evidence lies submerged and thus difficult, if not now impossible, to
recover, but submarine archaeological investigations in the
southwestern Pacific offer striking possibilities.

INTRODUCTION

When asked by the organisers of the Dalhousie conference on late
Quaternary sea-level correlation and its applications to address the
subject of archaeology and sea-level change in the southwestern
Pacific, this author could not fail to wonder, in the light of so much
already achieved, where he might start.
    An unlikely, but relevant, starting place was on the shores of
Golfo San Matias in northern Patagonia on the occasion of the post-
conference excursion of the last IGCP-200 meeting which was held in Mar
del Plata in Argentina in 1984. There, whilst studying the evolving
coastline of the Villarino Peninsula, the party examined a group of
shell middens and found a selection of stone tools on what had clearly
been a "stone-age" site. The very freshness of the site, however,
pointed to relatively recent occupation and this, in turn, was not
surprising when it is recalled that the first modern exploration of the
region took place only in the late 1880's. Until then, and for some
time later, this part of South America had been the domain of
aboriginal people who had not had more than limited and often hostile

195

D. B. Scott et al. (eds.), Late Quaternary Sea-Level Correlation and Applications, 195–206.
© 1989 by Kluwer Academic Publishers.

contact with the first immigrants of European extraction.

Contemplating the nature of the Patagonian coastal plain - dry, pebbly, scrub covered - it is not difficult to see why those Patagonian aboriginals had lived on the shores of Golfo San Matias. Shellfish existed in abundance and while the caloric value of such food is modest compared with that of reptiles, birds, mammals and even nuts, at least it was available as required all the year around - a veritable larder available to young and old, active and even infirm. The sites were "primitive" because the cultural level of the occupants was basically Stone Age: the tools representing not only the level of achievement of the inhabitants, but, as it happened, the best tools available for their needs. As immigration into the area by Argentinian colonisers advanced, the sites were abandoned by the aboriginals who, in a few short decades, all but died out, yet the evidence of their existence remains for us to examine today. That evidence reveals nothing of other, higher, levels of development which existed at the same time only a short distance away.

From this beginning, the writer began to consider the style of life of the indigenous peoples of other areas and in other times and more especially the effects of sea-level change on coastal communities wherever they may have occurred.

THE SOUTHWESTERN PACIFIC REGION AND THE LAST 10,000 YEARS

For the purpose of this paper the author has taken the region to be composed of the Malaysian peninsula, the Indonesian Archipelago, Papua-New Guinea and the islands of Melanesia, together with the continent of Australia (Figure 1). The time period covered is the Holocene and

Fig.1   The setting of the southwestern Pacific

through that time global sea level rose, in general terms, some 100 +
10 (or + 20)m with regional sea levels varying greatly (perhaps by more
than 50m) (Marcus & Newman 1983). Much of the rise was between 10,000
y B.P. and 6,000 or 5,000 y B.P. More significantly, parts of the
region are sites of active tectonism and volcanism, the consequence of
the Australo/Indian plate pushing northwards against the Eurasian and
the Pacific plates. It is a part of the globe which is characterised
by an abundance of islands.

The range of racial type and technical achievement of pre-European
communities found within this region exceeds that found anywhere else
on earth and presents the ethnographer with problems, many still
unsolved, about migrations and cultural evolution. Upon their arrival
in this vast region in the Sixteenth and later centuries, the European
discoverers were impressed by the ordered nature of life and the
evidence of prodigious feats of navigation amongst at least some of the
societies. On the other hand there were many parts, often only
sparsely populated, where life was, and remains, relatively simple.
Some societies of the latter type are still being discovered, as for
example, the Tasaday people of southern Mindanao Island of the
Philippines discovered as recently as June 1971.

In many cases, the European explorer and colonist managed to
record a great deal of information concerning the aboriginal peoples
before these original inhabitants, in some instances, either died out
or so changed their life style by adopting new ways and/or
interbreeding that little remained to identify them with their original
ways of life. Examples of this situation, significant in understanding
this region, were the aboriginal people of the Andaman Islands, and the
original Tasmanians.

Archaeologists and ethnographers have, particularly in the last
four decades, addressed the problem of origins and migrations. Their
work has been the product of painstaking research aided by the most
precise means of dating their finds. There is a vast bibliography
available (in particular Allen, J et al 1977; Bellwood, P 1978, and
Campbell, J 1984). Of particular interest has been the subject of the
migration and subsequent development of the often vastly differing
racial types into the region. As the work proceeds new discoveries are
made and these either confirm an existing theory or present a new and
confusing dimension. This is particularly true when dealing with dates
of occupancy of sites of domicile, (for example Groube, L et al 1986).
Throughout, one key aspect of the migration study has been the position
of sea level, particularly at the end of the Pleistocene, and its
subsequent rise.

Central to the whole area being considered is the position of sea
level in the southwestern Pacific region, in particular the
establishment of links between Sundaland and Sahul-land (Fig 2).
Sundaland comprises the southeast Asian landmass including Indonesia,
as far as Bali (and including Sulawesi, and the Philippines in some
definitions). Sahul-land comprises Australia and Papua New Guinea.
The two, however, are still separated by a broad transition zone known
as Wallacea. Early settlers, at the time of the last low-stand of sea
level or at the previous times of low-stand, would still have had to

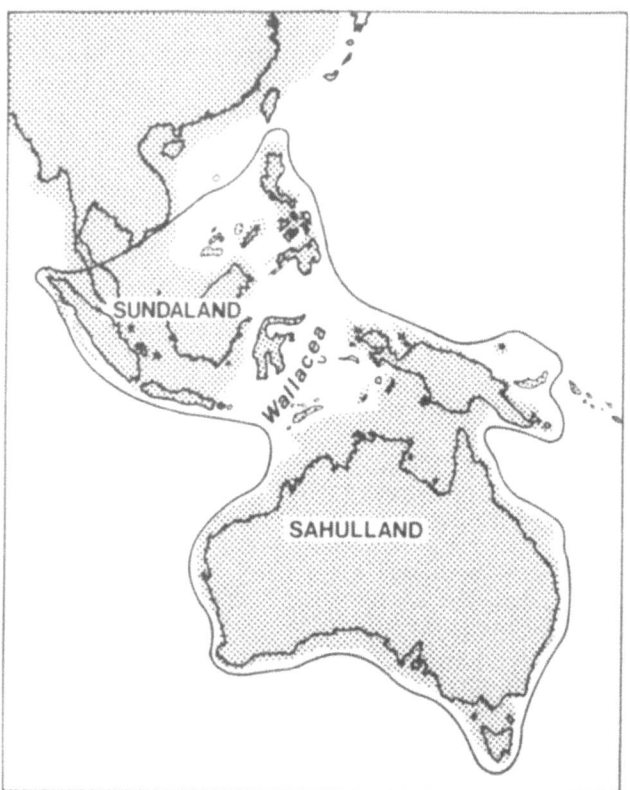

Fig.2 Sundaland, Sahul-land and Wallacea

migrate across several water gaps. A number of possible migration
routes across these water gaps have been suggested (Birdsell 1977 and
Fig 3). The discussions of the routes 'on offer' include an evaluation
of what could be seen of the 'stepping stones' by these early settlers
as they looked eastward from their contemporary sites. Evidence
(Birdsell, loc. cit) reveals that they would have been able to see
islands considerable distances away, depending on the elevation of the
observer and target, but it would still have required dire pressures to
trigger a migration toward an unknown goal. Why did an individual feel
pressed to go? Why should he persuade a group, his immediate family
group, to accompany him? What, in the end, prompted him and his
companions collectively to set forth? It had to be a group, for
without a group, colonisation could not be sustained. Anyway the group
was the essential basis of survival under any circumstance.

    Our evidence implies that the state of local contemporary
technical achievement was such that would-be colonisers did not step
into a carefully fabricated craft. Instead the voyage would either
have been by swimming the waters or, at most, propelling by hand a

simple bundle of wood and fibres which would become waterlogged within a short time. While there is little evidence to support a better style of water craft, evidence dictates against swimming, certainly swimming the final strait which by any route exceeds 60kms.

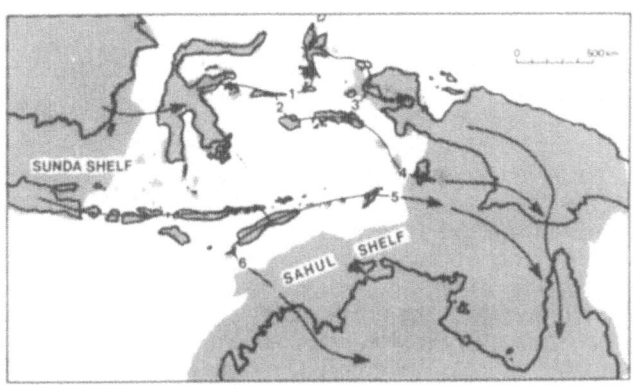

Fig.3 Possible routes and barriers

> *Barriers* *1. Sula to Obi : 93km*
> *2. Sula to Buru : 69km*
> *3. Seram to Misool Shelf : 68km*
> *4. Nuhu Tjut to Aru Is. Shelf : 103km*
> *5. Tanimbar to Aru Is. Shelf : 98km*
> *6. Roti to Sahul Shelf : 87km*

> *Distances allow for fall in sea level at last low stand of sea (ie. -100m OD).*
> *(Based on Birdsell, 1977).*

We can recognise that so long as our would-be voyager was not under pressure, there would be little reason other than curiosity to set out on a voyage. At times of low sea-level stand the coastal plain would, in places, provide the best and easiest terrain in which to live. Shell food from the shore, berries and fruits from the colonising forests, and game would have been abundant. He may have noted the migration of birds to and from the islands on the horizon, but there would have been little incentive for him to follow the birds. Climatic conditions in the region during global glacial advance would have been pleasant and very different from those facing settlers in the northern continents. One must suppose that it was not the opportunity of shortened distances at low sea level that called; it was, rather, the beginning of sea-level rise that pushed.

Barham (pers. comm.) reminds me that "sea-level change per se as a degree of sea-level change of 75% of the Holocene archaeological record is less than that observed in monthly tidal variation". This, of course, is undoubtedly true, but how does sea level rise? Surely this is a problem which faces all who work in this subject. What are we looking for? What is the event that our evidence records? Is it not true that barriers (beach bar, reef, vegetation, etc.) inhibit the effects of small even persistent rises? It is the catastrophic inundation that dramatically signals change and produces the evidence in the record - in such an area as the tectonically and volcanically active southwestern Pacific, there could have been many such signals.

As a succession of events led to the sea inundating easy territory two routes lay open to the inhabitants - landwards to the interior or, under dire threat, seawards to the land on the horizon.

EVOLUTION OF COLLECTIVE EXPRESSION

From our knowledge of early man from all parts of the world, we can assume that the easy living of the shore or anywhere else, for example, in the corridors of animal migration routes as were favoured sites in Eurasia, would have given man time to develop at least an element of leisure. Many archaeologists have calculated that once a group had enough to eat to permit some relief from the constant pressure to search for food, family, clan or tribal development could evolve a folk lore which expressed itself in story telling, art, and, for the want of an all embracing term, religion. In Europe, where for us the beginnings of art are recorded in the period of the last two ice advances, we can, without the written word, only surmise the evolution of legend. The tribal rites of today, the universal continuance of the tribal legend, should leave us in no doubt that these early inhabitants developed a collective identity. How else could the styles of their tools become so developed as to permit our archaeologists to ascribe them to cultures?

At the time of the onset of sea-level rise following the penultimate and last glacial maxima, however it was experienced, coastal dwelling peoples found themselves under threat - behind them a less hospitable land, peopled by perhaps unfriendly rivals - those who had aspired to the easy living of the coast but had been prohibited from it, or the outcasts of those who lived near the coast and who had for various tribal reasons been expelled and thus were certainly rivals and, perhaps, enemies.

We must assume that in the lore of the coastal peoples the islands of the horizon had been part of the backdrop of their life; in the face of new pressures where else was there to go but seawards.

All that appears here about the forces which dictated migration has a common thread with the many workers who have studied the migrations of the region. What we know is that sea level rose, (though admittedly in some areas the land has risen faster than the sea in response to tectonism), and that peoples of very different types and, in human terms, at widely separated times did migrate.

The two themes that I wish to develop are firstly, that archaeologists are all too often presented with the evidence of the outcast and, secondly that in fact there is a glaring gap in the data base since a considerable part of the record is submerged. Further, the latter, the submerged evidence, may be of a very different level of attainment. In every period the weaker are separated while the stronger take the better locations (Moseley, 1980).

Jones (1979) tells us of evidence pointing to at least two episodes of invasion of the Australian continent. The first colonisers, of archaic form, bearing the stamp of ancient Java Man, arrived more than 50,000 years ago while a more gracile Man arrived some 10 or 20,000 years later. Jones elegantly argues the case for the overlap of types and indeed the mixing of the gene pool to create the modern aboriginal of Australia. What is significant to any theme is that art, decoration and evidence of trying to come to terms with the mysteries of death are apparent from the earliest times, certainly from at least 32,000 years B.P. What is more the thoughts, as evidenced by art and funeral rites, are certainly equivalent to the thoughts of their Asian contemporaries. Several writers point to still other invasions by way of cultural leaps - the evidence the boomerang, barbed spears by 10,000 B.P. and the arrival of the dingo about 6,000 B.P. (Flood 1983) from the Indian sub-continent. With whom and by what means, we may well ask.

Here then is the ever present enigma. Man arrived in Sahul-land in large enough groups to colonise a continent in several distinct waves. But how? Swimming is unlikely, for individuals, even in desperation, can rarely swim more than 10 kilometres. Primitive water craft - see the well argued case of the short voyages from Tasmania after sea level rose - could not be expected to exceed that distance before becoming waterlogged yet by any route (see map Fig.3) at least 60km must be traversed. G G Simpson explained the enigma by the sweepstake principle, i.e. even the smallest chance can end in success if tried often enough. Even so individual successes will not be enough: fertile groups of at least half a dozen are required. Moreover, for success fire is important to increase the chance of survival.

THE LORE

It was at this stage the author naively imagined that there might be evidence in the legend or lore of island peoples.

The reason for this simplistic approach was based on the folk lore of the author's own youth for, as an Englishman brought up in Wales he was instilled with the story of the land of Cantref y Gwaelod. From today's coastlines three narrow ridges of pebbles protrude seawards into Cardigan Bay. These sarns, or, as translated, highways, were said to lead into the now submerged ancient lands which were inundated by the rising sea - the point of disaster being caused by a drunken watchman who failed to close the gates of the dykes against the rising tide on one night. Once flooded these lands, said to be comprised of delightful glades and flowing with milk and honey, could never be reclaimed. The legend is familiar Welsh folk lore. Subsequent marine work on the Bay has provided evidence of the Holocene sea-level rise with a major inundation about 5900 BP. The sarns, it should be recorded, are recognised as sea-modified glacial deposits.

Is this legend a long perpetuated story based on fact? In fairness it is only one of many such stories from Europe's Celtic fringe with the Atlantic Ocean. Legend, of course, may be involved in the Biblical story of Noah but when it comes to the legend of Atlantis most people and particularly the scientist balk at believing that story.

Notwithstanding his own scepticism, the author reviewed the legends of Oceania and Australasia. This proved to be fascinating reading thanks to the careful scholarship of Peter Bellwood (1978), Richard Carlyon (1981), Josephine Flood (1983), Joseph Campell (1984), Martha Beckwith (1985 – a reprint of the 1940 edition) and Donna Rosenberg (1986). The origin of Man and Woman – so very often from the sea – and indications of other lands across the sea are common themes. One has to be circumspect about the influence of European missionaries and the Biblical story. One, however, cannot fail to be impressed and to wonder about the themes expressed. Can Australian aboriginals really be recalling a time which preceded the extinction of the giant marsupials? Of a time before they developed weapons but once having learned that technology, drove the faunal giants to extinction? It fits, in broad terms, with some findings to which allusion has already been made (Jones 1979). If the legend has substance, then maybe so has the legend of arrival by boat, the search for water and subsequent migrations.

All too often the stories of the Gods of Oceania (Melanesia, Micronesia and of Australia), interesting as they are, leave one without a time dimension. Yet, legends are twinkling and unusually puzzling beacons. As one from a sea-faring race, the author is all too conscious of the lure of the wreckers lantern. Nevertheless there may be a way ahead here and an excursion towards the beacon is at least worthwhile. It will require on a researcher's part, all the skills which are applied to the conventional scientific approach.

What is the effect of migration, albeit forced, to reach the shores of another island, which is, in human terms, unoccupied? Is it a generous gift? If so, from whom and why? Is this a gift of Gods? There can be no doubt that the legend is embellished and there is often an opportunity in the retelling to enhance the stature of the teller and his affiliates. It may be hard, perhaps impossible, to separate fiction from fact. But legends do appear to represent a link between nature and man and thus began with the emergence of intellect and speech.

Today Polynesian groups 'play' with cosmic forces, yet Hawaiian stories are often limited to human activities on earth magnified by the incarnation of a divine ancestry. Admittedly, Polynesian and Hawaiian origins are later than those of the subject at issue. Yet, there is a route here for others to follow: a path between the absolute fact of science and the lore of the legend.

Hawaiian, Maori, Tahitian and even Japanese legends talk of 'wandering' islands, of an ancestral land sinking beneath the sea. Certainly until very recent times, and perhaps even still, it was the

practice of islanders of the South Seas when hard pressed to set sail
to a motherland never to return.   Even the Andamanese had legends of
people turning into birds and fishes to leave.

Nearly all ancient texts deal with the origin of Earth - with the
rolling back of the waters, letting dry land appear.   The Creation myth
of the Society Islands talks of "Taaroa, when there was no earth, no
sea, no sky and no man".

Scholars must remain sceptical for the issues are unclear.

THE RECORD

So far we have dealt with the spoken legend - flawed because it could
be embellished.   It, or they, only became a substantive record when
written down - something which occurred as recently as the middle of
this century in some cases.   There is, however, another record which we
can inspect for at least this was inscribed - art.

Cave paintings and frescos present modern man with images of what
early man thought.   These images are direct and the message is obvious.
They show in vivid detail the contemporary scene - particularly the
animals, weapons and the technology of hunting and aspects of what
might, in the broadest terms, be described as the rituals of religion.
The age range of such works is considerable.   If we can dismiss stories
of the creation, can we dismiss the similarities between cave art of
Europe and the rituals depicted in Australia?   Certainly by 18000 BP
rituals as depicted in art had crossed from Europe into Asia.   Art
forms were sufficiently revered and esteemed to be buried with the dead
- even dead children.   Styles of art, too, seemed to spread quickly.
The x-ray style of animal painting and engraving seems to have spread
rapidly at the time of the last low sea-level stand (Campbell, 1984) -
seemingly into Australia through Arnhem land.   The migration of ideas
must have been possible, but for this author two puzzles remain.
First, in no instances of cave art do we see drawings of water craft of
any sort.   Why?   Could they have been too ordinary?   Were they of no
significance to the artists?   The second is the omission of any
vegetation in such drawings.   Again was this because it was commonplace
and art, in its practiced form, was only concerned with the uncommon?

Here lies the next major enigma of early man's migration and the
evidence of driving forces - ideas could move, culture could move and
man could move but by what means?   Nowhere is the question posed more
vividly than in the southwestern Pacific.   Reducing sea level does not
remove the gaps.   Crossing the gaps is evidenced by the movement of
ideas.

In detective methodology one has to ask the three key questions
based on three words - motive, opportunity, method.   The motive may
just have been curiosity or, more seriously, the pressure of
population, perhaps in face of rising sea level and reduced resources,
the opportunity could have been a time of lower sea level, but this
might not have been of paramount importance.   The method remains
unidentified.

We are disposed to accept that Homo sapiens evolved from a common root. We know he could not have simply swum the distances involved. Larger creatures had succeeded in swimming - eg. the late Pleistocene pygmy elephant, others by rafting - eg. mice and rats. Diamond (1987) argues that the Komodo "dragon" in Wallacea evolved to prey on the elephant. The dragon was successful enough to drive the elephants to extinction yet survived until deer, pigs, buffalo, horses and goats were introduced by man by 5000 BP (sources Glover 1977 and Bellwood 1985) and in the millenia after that date. Putting aside the question of how the Komodo dragon survived after the demise of the elephant - most obviously other dragons and lesser mammalia - we are left with the evidence that a range of animals, many of a domesticated nature, had arrived by 5000 BP. Presumably these were transported by man. This in turn presupposes the ability to voyage - that is to control the direction of movement.

The great waves of migrations in the southwestern Pacific by navigators which today capture modern man's imagination only began after 4000 years ago, that is, about the time sea level stabilised near its present position and later. Today, all we can do is reflect on the major gap which exists in our record.

The archaeology of the presently submerged shelf has already attracted much attention. Much of this is vividly described by Flemming (see for example his 1985 work) and in Masters and Flemming (1983). Flemming has commenced a search of the NW Australian shelf on the Coolamundra Shoals. So far, pre-5000y BP evidence has not emerged. Pre-5000y BP discoveries promise to be hard to come by (Kraft et al. 1983) but it is not impossible that evidence may yet emerge as manned submersibles and remotely operated submarine vehicles become more widely used. The evidence we seek may yet be too ephemeral to survive or too deeply buried beneath younger sediments ever to be seen. Should evidence emerge it will fill two gaps - the first from the concealed 6% of our planet where to date only traces emerge (viz. the recent discovery of an artefact from the northern part of the North Sea), the second, a missing link in man's social evolution. The latter is of paramount importance for this missing link is, to mix metaphors, also the springboard of man's rapid advance.

CONCLUSION

The author makes no apologies for the inconclusive nature of his contribution: he remains convinced that the history of mankind in legend and in art may at least offer some leads. In conclusion, the author is heartened by the fact that the Woodward Chair in Natural History at Cambridge was endowed by Woodward with the express purpose of investigating the evidence for the Deluge. For nearly 250 years this Chair has been filled and Woodward's hopes unfilled.

A better understanding of the nature of sea-level change will help fill the archaeological gaps. A particularly key area is, without doubt, in the southwestern Pacific. Maybe there will just be time to produce the clues before the next sea-level rise; a time when the vast majority of the world's population, living as it does just above sea level, will be more concerned about their immediate future than their immediate past.

ACKNOWLEDGEMENTS

The author must thank the organisers of the Dalhousie conference for encouraging this excursion into the topic of this paper. Inconclusive as it must seem, the preparation of the paper permitted a fascinating insight into key issues facing anthropologists and archaeologists. Thanks, too, to many international authorities for their time and patience in advising the author. Finally acknowledgements to the reader who may not agree with the approach but as W Köppen told Alfred Wegener, "To work at subjects which fall outside the traditionally defined bounds of a science naturally exposes one to being regarded with mistrust by some, if not all those concerned, and being considered an outsider".

REFERENCES

Allen, J., Golson, J. & Jones, R. 1977. Sunda and Sahul. Academic Press London, 647p.

Beckwith, M. 1985. Hawaiian Mythology (reprinted from 1940 [1st edition]). University of Hawaii Press, Honolulu, 571pp.

Bellwood, P. 1978. Man's conquest of the Pacific. Collins, London, 462p.

Bellwood, P. 1985. Prehistory of the Indo-Malaysian archipelago. Academic Press New York.

Birdsell, J.D. 1977. Recalibration of a paradigm for the peopling of Greater Australia in Allen, J., Golson, J. & Jones, R. (eds) Sunda and Sahul. Academic Press London, pp.113-167.

Campbell, J. 1984. Historical atlas of world mythology Vol.1. Time Books Ltd, London, 303p.

Carlyon, R. 1981. A guide to the Gods. Heinemann, London, 402pp.

Diamond, J.M. 1987. Did Komodo dragons evolve to eat pygmy elephants? Nature, 326, p.832.

Flemming, N.C. 1985. Ice ages and human occupation of the continental shelf. Oceanus, 28, 18-25.

Flood, J. 1983. Archaeology of the Dreamtime. Collins, London, 288pp.

Glover, I.C. 1977. World Archaeology, **9**, 42-61.

Groube, L., Chappell, J., Muke, J. & Price, D. 1986. A 40,000 year old human occupation site at Huan Peninsula, Papua New Guinea. Nature, **324**, 453-455.

Jones, R. 1979. The fifth continent: problems concerning the human colonisation of Australia. Ann. Rev. Anthropol., **8**, 445-466.

Kraft, J.C., Belknap, D.F. & Kayan, I. 1983. Potentials for the discovery of human occupation sites on the continental shelf and near shore coastal zone, in Masters & Flemming (eds) 1983. Academic Press London, 87-120.

Marcus, L.F. & Newman, W.S. 1983. Hominid migrations and the eustatic sea level paradigm, in Masters & Flemming (eds) 1983. Academic Press London, 63-86.

Masters, P.M. & Flemming, N.C. (eds) 1983. Quaternary coastlines and marine archaeology. Academic Press London, 641pp.

Moseley, M.E. 1980. Settlement pattern determinants in the Viru and Moche Valleys. Unpublished pre-print for the Burg Wartenstein Symposium no.86: Prehistoric settlement pattern studies - Retrospect and Prospect. 18pp. Wenner-Gren Foundation for Anthropological Research, New York.

Rosenberg, D. 1984. World Mythology. Harrap, London, 520pp.

"In 1988 it was revealed by the Philippine authorities that the 'discovery' of the Tasaday people on southern Mindanao in 1971 was a hoax. Such a revelation does not affect the thesis of this paper".

SOME CONSIDERATIONS OF THE COMPILATION OF LATE QUATERNARY SEA LEVEL CURVES: A NORTH AMERICAN PERSPECTIVE

W. S. Newman
Department of Geology
Queens College of the City University of New York
Flushing, New York 11367
USA

R. R. Pardi
Department of Chemistry, Physics, and Environmental Science
William Paterson College of New Jersey
Wayne, New Jersey 07470
USA

R. W. Fairbridge
Department of Geological Sciences
Columbia University
New York, New York 10027
USA

ABSTRACT. A sea-level curve is either hand-drawn (as many of the earlier curves were) or the result of some statistical analysis (usually least-squares). The presentation of a given curve drawn through a set of empirical data points is a statement of some hypothesis or hypotheses. These hypotheses fall into three general categories;   1. that post-glacial sea level has risen smoothly and continuously along a trajectory which approximates a logarithmically or low-order polynomial curve intersecting the contemporary datum; 2. that post-glacial sea-level rise has been interrupted by one or more regressive episodes; 3. that the record of post-glacial sea level at any given locality is segmental so that segments of the curve may not project through the origin and thus may be recording local and regional neotectonic events.

Geoidal, neotectonic, and tidal range changes play crucial roles in varying sea-level trajectories through time. Our current investigations along the east coast of the United States demonstrate that intraplate seismotectonics are sometimes responsible for the confusing levels of basal peat dates.

D. B. Scott et al. (eds.), Late Quaternary Sea-Level Correlation and Applications, 207–228.

INTRODUCTION

Sea-level studies as a model scientific program

The literature on sea level contains a rich record of the interplay between field observation and theory. One aspect of that interplay has been the way in which empirical data have been presented. Every presentation within the vast literature on sea level has been a statement of some hypothesis related to the causes of relative sea-level change. The major (and many of the minor) causes of sea-level change were known and discussed (e.g. Daly, 1934) before they could be supported with empirical evidence; some fieldwork was driven by an attempt to confirm theory - some fieldwork drove theory. Insight into the more subtle causes of sea-level change was not driven exclusively by the accumulation of new data. Long before empirical observations demanded it, some of the potential causes of modulation of the simpler curves of isostasy and eustasy were well understood. Fairbridge (1961) enumerated several possible effects which could be superimposed on smoothly varying sea-level change, indicating these effects could be local rather than worldwide. Higgins (1965) summarized many of the potential effects on sea level, some of which should have magnitudes so small that they remain beyond our ability to detect today and probably for some time into the future.

The purpose of this paper is to review the development of the concept of post-glacial sea-level change, especially with respect to the interaction between hypotheses and the presentation of empirical information, with an admittedly North American viewpoint.

OVERVIEW OF THE COMPILATION OF SEA-LEVEL RECORDS

Sea level without chronology

Changes in elevation following the last glaciation were first displayed as simple isobase maps (DeGeer, 1888, Fennoscandia and Lake Bonneville; 1892, northeast North America) and as profiles of warped strandlines such as those reproduced in Daly (1934, Figure 52). These early maps were presented within only an approximate timeframe. Isobase maps have remained the traditional way to express the results of rheological models (Walcott, 1972a and b; Peltier, 1974; Cathles, 1975; Clark, 1980).

Sea level within the varve chronology framework

Nansen (1922) provided some of the earliest diagrams of sea level versus time (varve years) for the Fennoscandian area, as well as the earliest uplift rates versus time graphs (Figures 40 and 41 reproduced in Daly, 1934). These curves helped to direct attention to isostatic rebound but were not influential in quantifying the magnitude of sea-level changes outside the glaciated region. Sea-level change remained almost exclusively the domain of geophysicists until the advent of radiocarbon dating after 1947.

Early radiocarbon years - simple curves

Radiocarbon dating permitted the construction of a sea-level curve at any coastal location where samples appropriate for radiocarbon analysis were preserved, subject only to the availability of scientists with adequate resources. Initial efforts concentrated on those areas undergoing substantial isostatic rebound or dramatic transgression.

Isostasy in particular, and eustasy to a lesser extent are gradual processes. There are theoretical limits (Walcott, 1972a and b; Cathles, 1975; Peltier, 1980; Farrell and Clark, 1976; Clark, 1980) to the rate at which the Earth can respond to the redistribution of mass just are there are limits, though of shorter period, to the rate of change of ocean volume. Fully aware of these limits, and fully aware of the limits of the data base, early workers tended to portray sea-level change as a smoothly varying curve or straight line.

Shepard (1963), skeptical of the quality of many of the then current radiocarbon determinations, hand drew a continuously transgressive curve through the available world sea-level data. That curve was and still is extensively used as a reference in discussions of alternative sea-level curves. Emery and Garrison (1967) felt that the scarcity of data did not justify drawing anything but a least-squares straight line (all the sea-level diagrams in that paper are either single straight lines or segmented straight lines) through sea-level data points for the Texas and Atlantic coastal regions. They did feel, however, that the data were of sufficient quality to caution against accepting any local sea-level curve as representative of worldwide absolute changes in sea level. Redfield (1967) used an involved scheme to attempt to remove the effects of local subsidence or upwarping on sea-level curves (his Figures 1 and 3) but resorted to hand-drawn straight-line segments to support his claim for variable but continuous post-glacial sea-level rise.

The implication in these presentations was that the error in sampling and dating sea-level indicators obscured the simple, underlying eustatic signal. The stage was set for the search for the "universal", eustatic sea-level curve, if only the "perfect" locale and samples could be found.

Sea-level curves become complex

As the number of radiocarbon dates on sea-level indicators increased it became possible to define, with some confidence, structure within the overall post-glacial eustatic and isostatic curves. Curray (1961) attempted to show the synchroneity of sea-level fluctuations with climatic fluctuations on land for an area (the Texas Gulf coast) where shelf features have not been above present sea level for at least 30,000 years. Fairbridge (1961) first suggested a sequence of higher-than-present Holocene sea-level events, refuting the concept of stable post-6000 yBP sea level (Fisk and McFarlan, 1955; Godwin et al., 1958; Gould and McFarlan, 1959; McFarlan, 1961). Subsequent investigations either supported Fairbridge (Van Andel and Laborel, 1964; Bloch, 1965; Hume, 1965; Hafemann, 1965; Behrens, 1966; and Lind, 1969); argued for continuously lower-than-present sea level (Scholl, 1964; Scholl and Stuiver, 1967a and b); or offered further support for a stable post-6000 yBP sea level (McIntire and Morgan, 1964; Redfield and Rubin, 1972; and Coleman and Smith, 1964). Figures of sea-level curves such as those of Fairbridge (1961) and Morner (1976) were often overlapped (such as in Morner, 1981) to show either the synchroneity of a number of curves or the lack thereof.

If historical changes in sea level are in any way congruent with the longer, post-glacial changes, then studies of tide-gauge stations worldwide (Gutenberg, 1941; Fairbridge and Krebs, 1962; and Lisitzin, 1974) would reject any model of sea-level stability. The presentation of historical tide-gauge records substantially paralleled the presentation of post-glacial sea-level studies.

As it became clearer that complexity within sea-level records was the norm rather than the exception, attempts were made to employ surface (distance-distance-elevation) mapping (isobases based on a radiocarbon chronology rather than stratigraphic markers). McCann and Chorley (1967) employed trend-surface maps to illustrate sea-level change in western Scotland, while Flemming (1978) used trend-surface maps to show the extent of eustasy and tectonism within the northeast Mediterranean. Beginning with their 1978 paper Newman et al. (1980b, 1981, 1984) explored various surface-generating schemes to portray local, regional and worldwide trends in sea level through time and space.

End of the search for the "universal", eustatic curve

Even before the empiricists began struggling with the complexities of non-uniform sea-level change, the geophysical modelers were resurrecting some of the long-forgotten explanations for such complexity. Walcott (1972a), referring back to Daly (1925), thoroughly dismissed any concept of coastal sea-level stability. In that paper Walcott presented a schematic summary of worldwide sea-level trends (his Figure 6) very similar to one presented earlier by Newman (1968). One diagram in particular, which appeared a few years later, has come to symbolize our current approach to sea-level studies - the scatter plot (Newman et al., 1981; Marcus and Newman, 1983) showing all the known radiocarbon-dated sea-level indicators as a function of age and elevation (Figure 1).

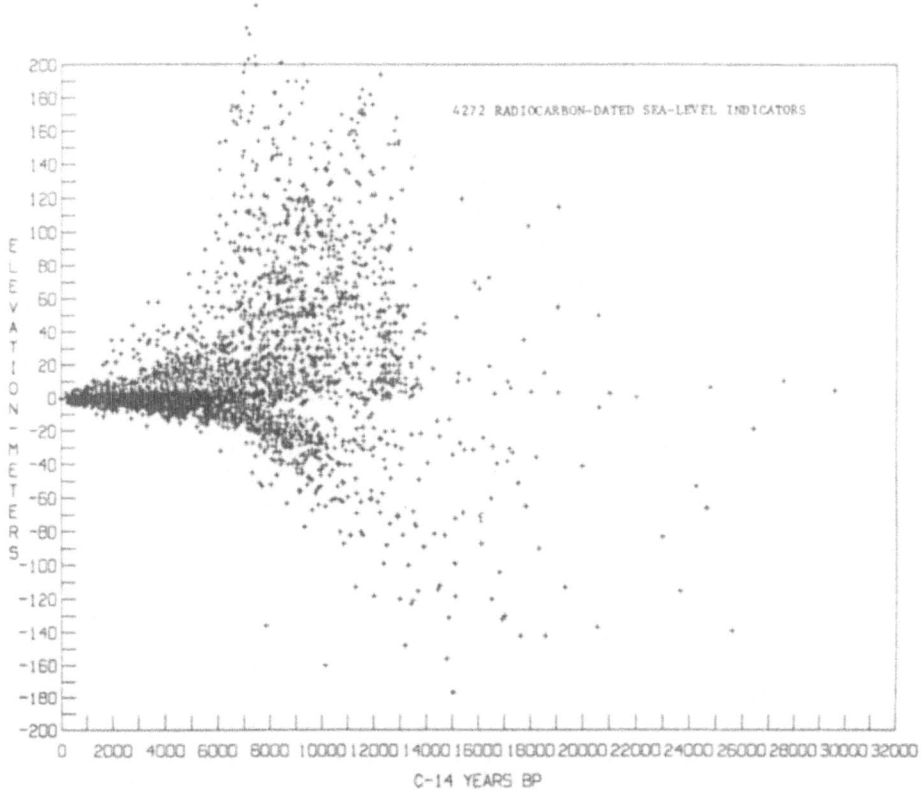

FIGURE 1: Scatter diagram of world sea-level data (Figure 2 from Newman et al., 1984). Plotted here as a function of age and depth are 4272 worldwide sea-level indicators without selection with respect to area.

curves. Curray (1961) attempted to show the synchroneity of sea-level fluctuations with climatic fluctuations on land for an area (the Texas Gulf coast) where shelf features have not been above present sea level for at least 30,000 years. Fairbridge (1961) first suggested a sequence of higher-than-present Holocene sea-level events, refuting the concept of stable post-6000 yBP sea level (Fisk and McFarlan, 1955; Godwin et al., 1958; Gould and McFarlan, 1959; McFarlan, 1961). Subsequent investigations either supported Fairbridge (Van Andel and Laborel, 1964; Bloch, 1965; Hume, 1965; Hafemann, 1965; Behrens, 1966; and Lind, 1969); argued for continuously lower-than-present sea level (Scholl, 1964; Scholl and Stuiver, 1967a and b); or offered further support for a stable post-6000 yBP sea level (McIntire and Morgan, 1964; Redfield and Rubin, 1972; and Coleman and Smith, 1964). Figures of sea-level curves such as those of Fairbridge (1961) and Morner (1976) were often overlapped (such as in Morner, 1981) to show either the synchroneity of a number of curves or the lack thereof.

If historical changes in sea level are in any way congruent with the longer, post-glacial changes, then studies of tide-gauge stations worldwide (Gutenberg, 1941; Fairbridge and Krebs, 1962; and Lisitzin, 1974) would reject any model of sea-level stability. The presentation of historical tide-gauge records substantially paralleled the presentation of post-glacial sea-level studies.

As it became clearer that complexity within sea-level records was the norm rather than the exception, attempts were made to employ surface (distance-distance-elevation) mapping (isobases based on a radiocarbon chronology rather than stratigraphic markers). McCann and Chorley (1967) employed trend-surface maps to illustrate sea-level change in western Scotland, while Flemming (1978) used trend-surface maps to show the extent of eustasy and tectonism within the northeast Mediterranean. Beginning with their 1978 paper Newman et al. (1980b, 1981, 1984) explored various surface-generating schemes to portray local, regional and worldwide trends in sea level through time and space.

End of the search for the "universal", eustatic curve

Even before the empiricists began struggling with the complexities of non-uniform sea-level change, the geophysical modelers were resurrecting some of the long-forgotten explanations for such complexity. Walcott (1972a), referring back to Daly (1925), thoroughly dismissed any concept of coastal sea-level stability. In that paper Walcott presented a schematic summary of worldwide sea-level trends (his Figure 6) very similar to one presented earlier by Newman (1968). One diagram in particular, which appeared a few years later, has come to symbolize our current approach to sea-level studies - the scatter plot (Newman et al., 1981; Marcus and Newman, 1983) showing all the known radiocarbon-dated sea-level indicators as a function of age and elevation (Figure 1).

The end of the search for a locale where the record of sea level would be purely eustatic opens the way for sea-level studies to explore what is essentially the inverse problem - namely, how can we use the differences in sea-level curves to tell us something more than just the record of changing sea level. Faulting events along coastal reaches can be expected to produce stepped discontinuities in the sea-level record (Huber, 1975). Such discontinuities have been reported from a number of locations worldwide. Taira (1975) used offsets in the sea-level record for the Hengchun Peninsula of Taiwan as evidence for faulting with throws of up to 10 meters for one episode. Vita-Finzi (1975) observed similar offsets of relict beaches of the southern Iranian coast, indicating repeated uplifts.

As discussed by Newman (1980c, 1987), faulting will be reflected within a sea-level diagram by "steps" within the curve - segments of the curve will not project through the origin as straight lines or curves (Figure 2). Though this can be easily illustrated in conventional time-elevation plots, faulting events are much more difficult to illustrate using surface plotting techniques.

INTERACTION OF PRESENTATION AND THEORY: EXAMPLE, THE U.S. EAST COAST

By of the number of studies and the quantity of data collected, the northeast coast of North America has the most intensely examined record of sea level in the world. While this fact is in part due to human geography, the northeast North American coast also exhibits among the world's most varied responses to sea-level change. So varied is that response that the sea-level record of the eastern shore of North America has become the testing ground for many models of sea-level change. Thus, Bloom (1967) looked to the sea-level record of the eastern shore of North America to support the concept of hydroisostasy. Yet sea level along that stretch of coast remains a hotly debated issue.

Abandoning any concept of Holocene coastal stability, Belknap and Kraft (1977) proposed that the outer shelf and shelf edge of the Delaware/Maryland coast has downwarped as much as 40 meters in 10,000 years. Dillon and Oldale (1978) presented evidence from the continental shelf off southern New England for a complex of tilting (NE) and broad warping (striking SW from the central New Jersey coast). MacIntyre et al. (1978) questioned the validity of the use of shelf shell dates to support sea-level chronologies, but their conclusion was refuted by the subsequent study of Field et al. (1979) whose shelf peat data closely approximated the Belknap and Kraft curve. That the absolute magnitude of the post-glacial transgression was less than 100 m. was suggested by the work of Blackwelder et al. (1979), while Oldale and O'Hara (1980) placed sea level at -70 m at 12,000 yBP off the southern coast of New England. Finally, Blackwelder (1980) could find no evidence for warping of the continental shelf between New York and South Carolina while placing sea level not much below -30 m at 12,000 yBP. The sea-level record

214

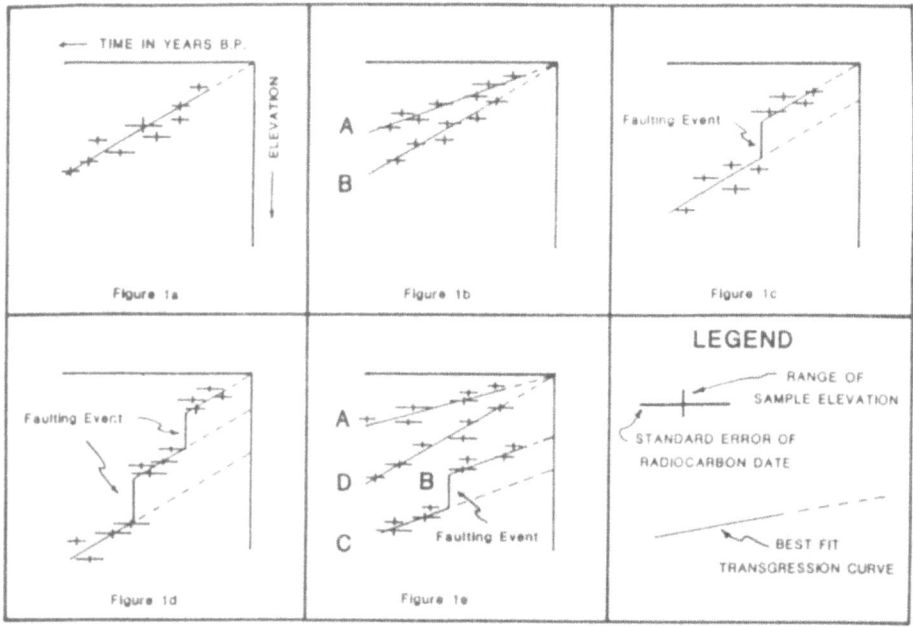

FIGURE 2: Illustration of the effect of faulting on sea-level curves (Figure 1 from Newman et al., 1987). These figures were meant to show how faulting events would be reflected in sea-level curves at specific locales.

north of the Cape Fear Arch is distinctly different from that south of the Arch (Cinquemani, et al., 1982). Studies of historical tidal records (Hicks et al., 1983) and repeated precise leveling (Brown, 1978) confirm that this dichotomy continues to the present day. In both the above studies, the authors chose to use isobase maps to summarize the extent and sign of vertical crustal movement.

Faced with such a confusing set of claims, Cinquemani et al. (1982) attempted to resolve the entire dataset by taking the approach of plotting the data as a continuous surface rather than discrete curves. Surface-generating techniques have strengths and weaknesses; these are discussed in detail in Pardi and Newman (1987). Once the data set was examined within millenial cohorts, at least some of the dichotomies listed above were resolved (Figure 3). The inherent problem in the use of isobase-type contour diagrams to present sea-level data along a coast is that programs to generate such closed-contour projections project trends into geographical areas where data do not exist. Such projections may be relatively accurate in regions with complex coastal geography, but they are almost certainly invalid along straight coastlines such as that of the northeast coast of North America. Cinquemani et al. (1982) attempted to resolve part of this problem by leaving contours open in such presentations (Figure 4).

Such presentations did, however, permit a more open view of sea-level variations. Free of the confines of a constant or constantly transgressive Holocene sea-level, interdisciplinary studies, especially those involved with archeology (Brooks et al., 1979; Colquhoun et al., 1980; and DePratter and Howard, 1977), have explored new paradigms within the framework of fluctuating sea level.

While the study of past sea level is a subject of fundamental interest in and of itself, it has been from its infancy the handmaiden of geophysics. Isostatic rebound is the physician's hammer with which the Earth has conveniently been struck, affording geophysicists the opportunity to observe the response of the Earth to a rapid, deforming insult. Many geophysical studies have concentrated on the area of central uplift, but one particular problem derives more from the edge of the glaciated region, namely the search for "peripheral bulge collapse". Daly (1934) recognized that the mode of response of the surface of the Earth would be particularly sensitive to the rheological characteristics of the upper mantle. While there is some latitude in model parameters with respect to the central uplift, one area is particularly sensitive to variations in model parameters - the area immediately peripheral to the ice cover. Depending primarily on the viscosity and viscosity distribution of the Earth's mantle, the area peripheral to the central ice mass will either be depressed, remain stable or bulge upward while the central mass is depressed isostatically.

216

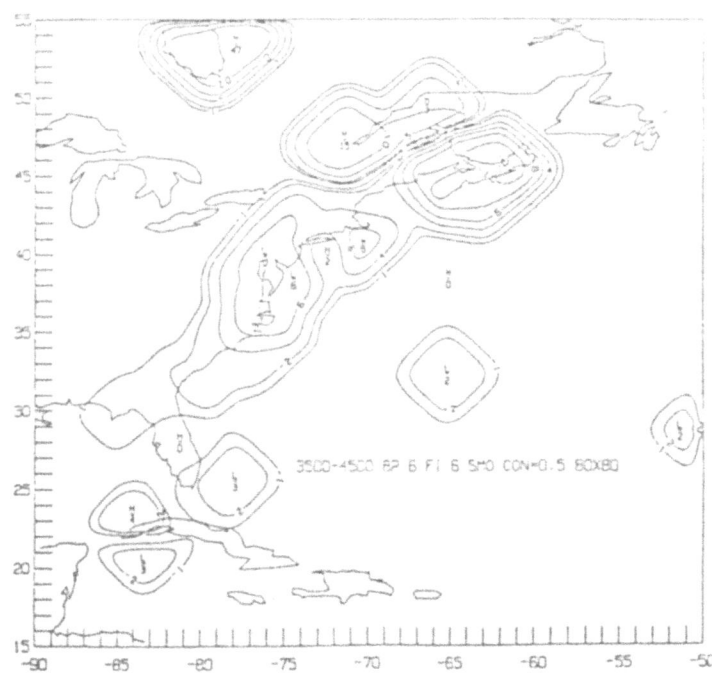

FIGURE 3: Isobase map for the east coast of North America 3500 to 4500 yBP (Figure 4 from Cinquemani et al., 1982). A contour generating program was applied to all points falling within the age limits and the geographic area of the map. The contours are, then, isobases of elevation or depth for specific periods of time.

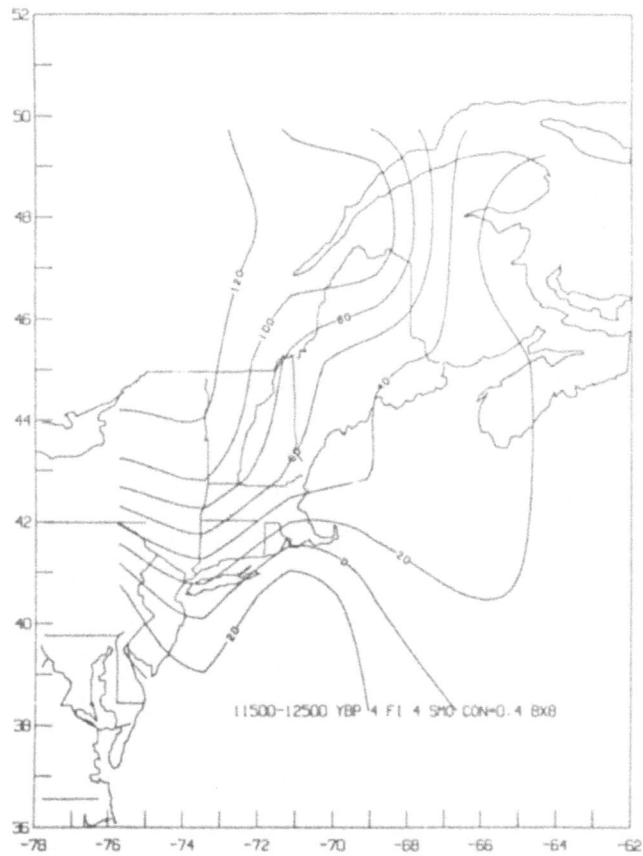

FIGURE 4: Isobase map for the east coast of North America 11500 to 12500 yBP (Figure 3 from Newman et al., 1987). The presentation is meant to convey roughly the same type of information as shown in Figure 4, however, here the contour-generating program was not constrained to close or terminate the contour lines - thereby giving at least some impression of the uncertainty in projecting isobases away from coastal areas.

Here is an area that amply illustrates the interaction
between theory, fieldwork and the presentation of data. Long after
Daly first proposed the existence of a "bulge" peripheral to the
central uplift, Newman and Rusnak (1965) and Newman et al.
(1971) proposed to have found evidence for it in the sea-level record off
Virginia and the New York area. Around the same time Walcott (1972a)
was troubled by the patchy nature of sea-level data even in this, the
most thoroughly-sampled area in the world. Cathles (1975) remained
unconvinced that the data proved the existence of a collapsing bulge
and he modified his rheological model accordingly. There followed,
however, a mounting tide of field evidence (e.g., Scott and Medioli,
1982; Quinlan and Beaumont, 1981) which supported the concept of a
collapsing Late Quaternary Early Holocene forebulge and which forced
a reassessment of the theoretical models. Pardi and Newman (1987)
have joined the argument in favor of a forebulge with a method for
the simultaneous presentation of the entire, sea-level data set for
the east coast of North America. While all the sea-level data seen at
one time (Figure 5 here, Figure 6 in Newman et al., 1984) fill both
geographical and time-elevation space, individual curves at specific
locales more often than not contain substantial gaps. As noted above,
the initial approach to this problem of discontinuous data made use
of trend-surface maps. However, as discussed by Pardi and Newman
(1987) surface maps based exclusively on coastal data can be
misleading, and they proposed an alternative-space map
(distance-elevation-time) as one way of exploring large areas and
numbers of data points along more or less straight coasts (Figures 5,
east coast of North America, and 6, the St. Lawrence valley from
Montreal to Newfoundland). Details of this approach are described in
Pardi and Newman (1987) - the figures can be viewed as essentially a
"stack" of cardlike sea-level curves forming a surface rather than an
line. Recently, Officer et al. (1987) have taken that presentation
(shown in Figure 5) and used it as a basis for refining the
rheological model to fit the observed coastal response, lending
renewed, theoretical support to the concept of peripheral bulge
collapse.

DISCUSSION AND CONCLUSION

The presentation of sea-level data faces both the intrinsic
limitations of dimensionality as well as the consequences of inherent
error. Two dimensional plots of elevation and distance will likely
remain a standard tool for evaluating sea-level changes within
circumscribed areas. Neither the distance-distance-elevation
(isobase) (Cinquemani et al., 1982; Newman, et al., 1981, 1980b,
1984) nor the distance-time-elevation (Petersen, 1985; Pardi and
Newman, 1987) presentation of changing sea level can totally resolve
all of the complexities in the data. What they can do is show where
data deserts are, where the most "sensitive" areas are, and where
there are inconsistencies in the data, by providing a method of
looking at all or a significant portion of the now massive sea-level

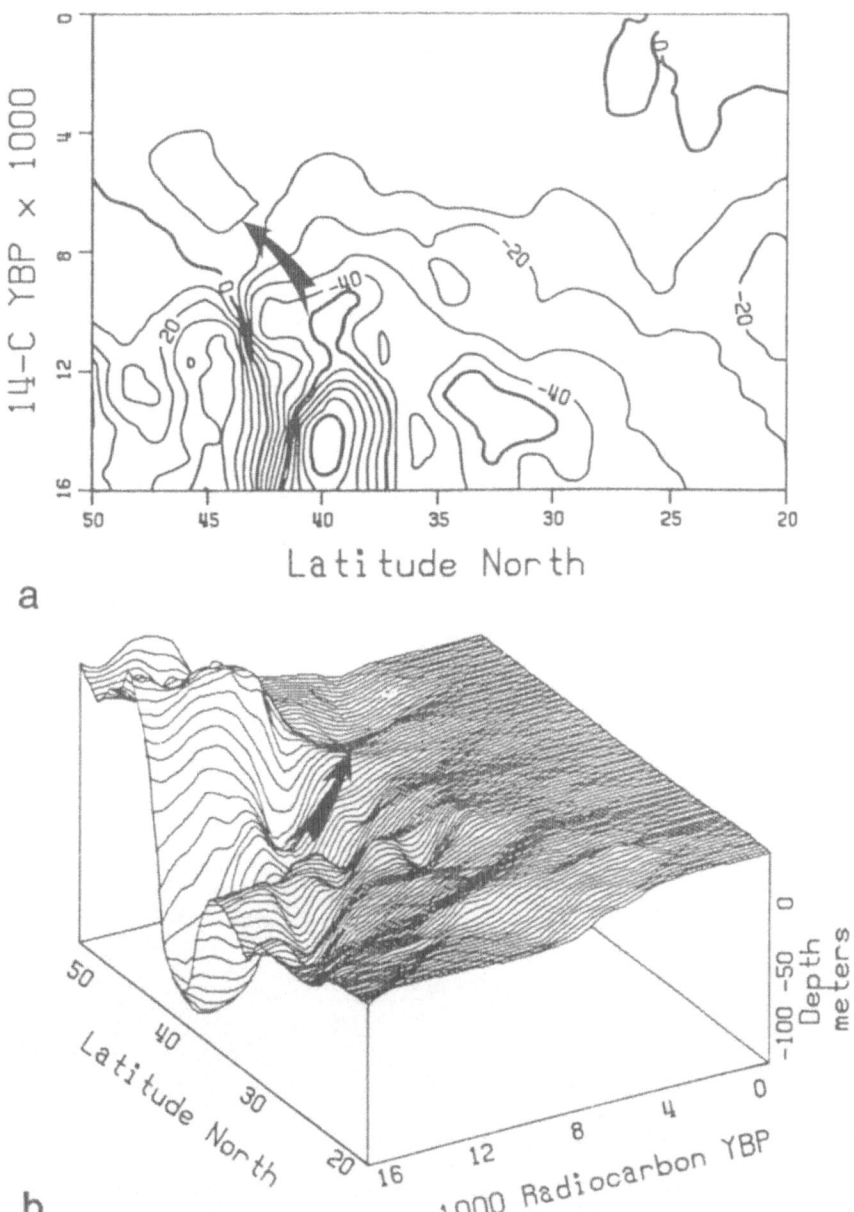

FIGURE 5: Contour and 3D diagram of sea-level data from east coast of North America (Figure 3 from Pardi and Newman, 1987). See that reference for a discussion of the methods and assumptions used in generating such diagrams (also for Figure 6). The arrow here suggests the possible path of a collapsing peripheral bulge through time and space.

220

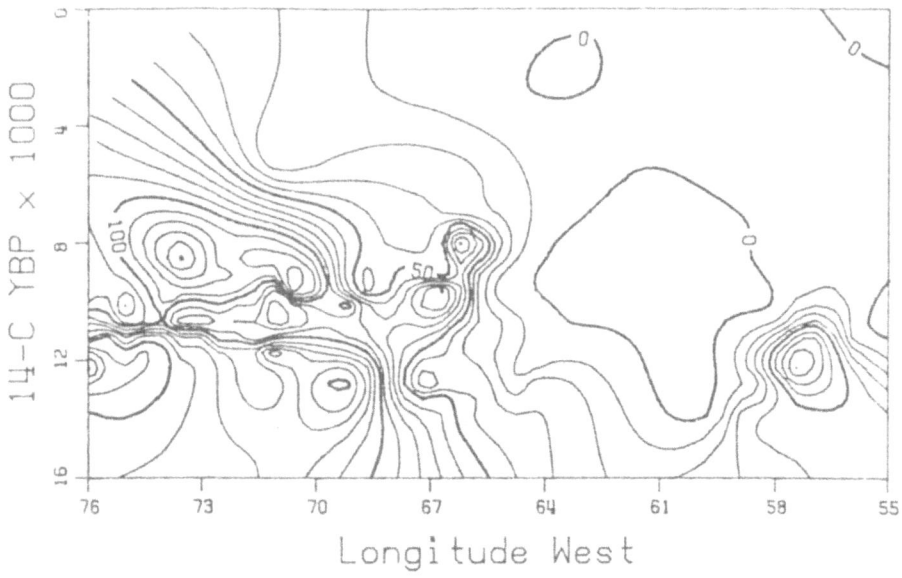

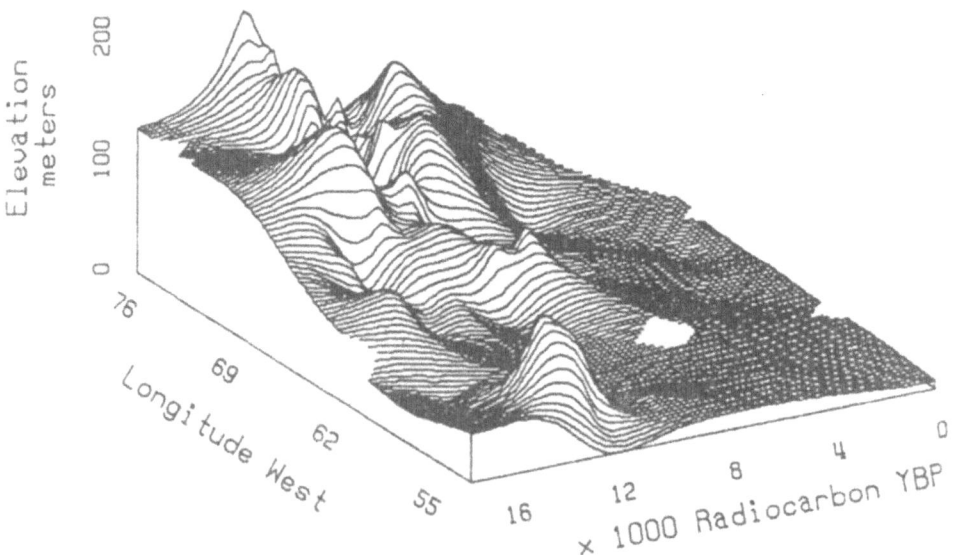

FIGURE 6: Contour and 3D diagram of sea-level data from St. Lawrence River Valley to Newfoundland (this paper)

data base.

When looking at massive amounts of data we must never lose sight of the individual data point nor the value of the individual field study. Newman's compilation of over 4500 sea-level dates was not simply a cataloging task. He had made decisions about the validity and meaning of each and every entry. He very likely knew the investigator personally and discussed the selection of samples in detail. He very likely had visited the sample collection site or even participated in the sampling for many of the North American sites. He was very aware of the impact that various methods of presentation and analysis could have on the process of discovery. Let us hope that the same rigor is applied to future sea-level research and that the interaction of presentation with theory and fieldwork will continue to lead to a clarification of post-glacial sea-level changes.

REFERENCES

Behrens, E.W., 1966, Recent emerged beach in eastern Mexico: Science, v. 152, p. 642-643.

Belknap, D.F., and Kraft, J.C., 1977, Holocene relative sea-level changes and coastal stratigraphic units on the northwest flank of the Baltimore Canyon Trough Geosyncline: Journal of Sedimentary Petrology, v. 47, p. 610-629.

Blackwelder, B.W., Pilkey, O.H., and Howard, J.D., 1979, Late Wisconsinan sea levels on the southeast U.S. Atlantic Shelf based on in-place shoreline indicators: Science, v. 204, p. 618-620.

Blackwelder, B.W., 1980, Late Wisconsinan and Holocene tectonic stability of the United States mid-Atlantic coastal region: Geology, v. 8, p. 534-537.

Bloch, M.R., 1965, A hypothesis for the change of ocean levels depending on the albedo of the polar ice caps: Paleogeography, Paleoclimatology, Paleoecology, v. 1, p. 127-142.

Bloom, A.L., 1967, Pleistocene shorelines: A new test of isostasy: Geological Society of America Bulletin, v. 78, p. 1477-1494.

Brooks, M.J., Colquhoun, D.J., Pardi, R.R., Newman, W.S., and Abbott, W.H., 1979, Preliminary archeological and geological evidence for Holocene sea level fluctuations in the Lower Cooper River valley, South Carolina: The Florida Anthropologist, v. 32, no. 3, p. 85-103.

Brown, L.D., 1978, Recent vertical crustal movement along the east coast of the United States: Tectonophysics, v. 44, p. 205-231.

Cathles, L.M., III, 1975, The Viscosity of the Earth's Mantle: Princeton University Press, Princeton, 386 pp.

Cinquemani, L.J., Newman, W.S., Sperling, J.A., Marcus, L.F., and Pardi, R.R., 1982, Holocene sea-level changes and vertical movements along the east coast of the United States: a preliminary report: p. 13-33 in: D.J. Colquhoun, ed., Holocene Sea Level Fluctuations, Magnitude and Causes, Department of Geology, University of South Carolina, Columbia, South Carolina.

Clark, J.A., 1980, A numerical model of worldwide sea-level changes on a viscoelastic earth: in: N.-A. Morner, ed., Earth Rheology, Isostasy and Eustasy, Wiley, New York, 599 pp.

Coleman. J.M., and Smith, W.G., 1964, Late Recent rise of sea level: Geological Society of America Bulletin, v. 75, p. 833-840.

Colquhoun, D.J., Brooks, M.J., Abbott, W.H., Stapor, F.W., Newman, W.S., and Pardi, R.R., 1980, Principles and problems in establishing a Holocene sea-level curve for South Carolina: in: J.D. Howard, C.B. DePratter, and R.W. Frey, eds, Excursions in Southeastern Geology, Geological Society of America Guidebook, v. 20, p. 143-159.

Curray, J.R., 1961, Late Quaternary sea level: a discussion: Geological Society of America Bulletin, v. 72, p. 1707-1712.

Daly, R.A., 1925, Pleistocene changes of level: American Journal of Science, series 5, v. 10, p. 281-313.

Daly, R.A., 1934, The Changing World of the Ice Age: Yale University Press, New Haven, 271 pp.

De Geer, G., 1888 and 1890, Om skandinaviens nivaforandringar under Qvartarperioden: Geol. Foren. Stockh. Forhandl., 10 and 12, p. 366-379 and 61-110.

De Geer, G., 1892, On Pleistocene changes of level in eastern North America: Proceedings of the Boston Society of Natural History, v. 25, p. 454-477.

DePratter, C.B., and Howard, J.D., 1977, History of shoreline changes determined by archeological dating: Georgia Coast U.S.A.: Transactions of the Gulf Coast Association of Geological Societies, v. 27, p. 252-258.

Dillon, W.P., and Oldale, R.N., 1978, Late Quaternary sea-level curve: Re-interpretation based on glacio-eustatic influence: Geology, v. 9, p. 56-60.

Emery, K.O., and Garrison, L.E., 1967, Sea levels 7,000 to 20,000 years: Science, v. 157, p. 684-687.

Fairbridge, R.W., and Krebs, O.A. Jr., 1962, Sea level and the southern oscillation: Geophysical Journal of the Royal Astronomical Society, v. 6, p. 532.

Fairbridge, R.W., 1961, Eustatic changes in sea level: p. 99-185 in: L.H. Ahrens, F. Press, K. Rankama, and S.K. Runcorn, eds., Physics and Chemistry of the Earth, v. 4, Pergamon Press, New York, 317 pp.

Fairbridge, R.W., 1962, World sea-level and climatic changes: Quaternaria, v. 6, p. 111-134.

Farrell, W.E., and Clark, J.A., 1976, On postglacial sea level: Geophysical Journal of the Royal Astronomical Society, v. 46, p. 647–667.

Field, M.E., Meisburger, E.P., Stanley, E.A., and Williams, S.J., 1979, Upper Quaternary peat deposits on the Atlantic inner shelf of the United States: Geological Society of America Bulletin, Part I, v. 90, no. 7, p. 618–628.

Fisk, H.N., and McFarlan, R., Jr., 1955, Late Quaternary deltaic deposits of the Mississippi River: Geological Society of America Bulletin Special Paper 62, p. 279–302.

Flemming, N.C., 1978, Holocene eustatic changes and coastal tectonics in the northeast Mediterranean: implications for models of coastal consumption: Philosophical Transactions of the Royal Society of London A, v. 289, no. 1362, p. 405–458.

Godwin, H., Suggate, R.P., and Willis, E.H., 1958, Radiocarbon dating of the eustatic rise in ocean-level: Nature, v. 181, p. 1518–1519.

Gould, H.R., and McFarlan, E., 1959, Geologic history of the chernier plain, Southwestern Louisiana: Transactions of the Gulf Coast Association of Geological Societies, v. 9, p. 1–10.

Gutenberg, B., 1941, Changes in sea level, postglacial uplift, and mobility of the earth's interior: Geological Society of America Bulletin, v. 52, p. 721–772.

Hafemann, D., 1965, Die niveauveranderungen an den Kusten Kretas seit dem Altertum: Wiesbaden, Franz Steiner Verlag, 84 pp.

Hicks, S.D., DeBaugh, H.A., and Hickman, L.E., 1983, Sea Level Variation for the United States, 1855-1980: National Ocean Service, Rockville, Maryland.

Higgins, C.G., 1965, Causes of relative sea-level changes: American Scientist, v. 53, p. 464–476.

Huber, N.K., 1975, Marine terraces, datum planes for study of structural deformation: Earthquake Information, v. 7, no. 3, p. 3–7.

Hume, J.D., 1965, Sea-level changes during the past 2,000 years at Point Barrow, Alaska: Science, v. 150, p. 1165–1166.

Lind, A.O., 1969, Coastal landforms of Cat Island, Bahamas: Department of Geography, University of Chicago, Research Paper No. 122, 156 pp.

Lisitzin, E., 1974, Sea Level Changes: Elsevier, New York, 286 pp.

MacIntyre, I.G., Pilkey, O.H., and Stuckenrath, R., 1978, Relict oysters on the United States Atlantic continental shelf: A reconsideration of their usefulness in understanding late Quaternary sea-level history: Geological Society of America Bulletin, v. 89, p. 277-282.

Marcus, L.F., and Newman, W.S., 1983, Hominid migrations and the eustatic sea-level paradigm: a critque: p. 63-85 in: P.M. Masters and N.C. Flemming, eds., Quaternary Coastlines and Marine Archaeology, Academic Press, 641 pp.

McCann, S.B., and Chorley, R.J., 1967, Trend surface mapping of raised shorelines: Nature, v. 215, p. 611-612.

McFarlan, E., Jr., 1961, Radiocarbon dating of Late Quaternary deposits, South Louisiana: Geological Society of America Bulletin, v. 72, p. 129-158.

McIntire, W.G., and Morgan, J.P., 1964, Recent geomorphic history of Plum Island, Massachusetts and adjacent coasts: Louisiana State University, Coastal Studies Series, no. 8, 44 pp.

Morner, N.-A., 1976, Eustasy and geoidal changes: Journal of Geology, v. 84, p. 123-151.

Morner, N.-A., 1981, Holocene sea level fluctuations on a global scale: evidence for geoidal eustasy: abstracts, IGCP-61, South Carolina.

Nansen, F., 1922, The strandflat and isostasy; Norske Vid. Akad. Oslo, Ser. Mat. Nat. Sci., v. 11, p. 1-350.

Newman, W.S., and Rusnak, G.A., 1965, Holocene submergence curve of the eastern shore of Virginia: Science, v. 148, p. 1464-1466.

Newman, W.S., 1968, Coastal stability: p. 150-156 in: R.W. Fairbridge, ed., The Encyclopedia of Geomorphology, Reinhold, New York.

Newman, W.S., Marcus, L.F., and Pardi, R.R., 1981, Paleogeodesy: late Quaternary geoidal configurations as determined by ancient sea levels: in: I. Allison, ed., Sea Level, Ice, and Climatic Change, International Association of Hydrological Sciences, Publication No. 131, Washington, D.C., p. 263-275.

Newman, W.S., Fairbridge, R.W., and March, S., 1971, Marginal subsidence of glaciated areas: United States, Baltic and North Seas: p. 795-801, in: M. Ters, ed., Etudes sur le Quaternaire dans le Monde, Union Internationale pour l'Etude du Quaternaire, Paris.

Newman, W.S, Cinquemani, L.J., Pardi, R.R., and Marcus, L.F., 1980a, Holocene delevelling of the United States East Coast: p. 449-463 in: N.-A. Morner, ed., Earth Rheology, Isostasy, and Eustasy, Wiley, New York, 599 pp.

Newman, W.S., Marcus, L.F., Pardi, R.R., Paccione, J.A., and Tomecek, S.M., 1980b, Eustasy and deformation of the geoid: 1,000-6,000 radiocarbon years B.P: p. 555-567, in: N.-A. Morner, ed., Earth Rheology, Isostasy and Eustasy, Wiley, New York, 599 pp.

Newman, W.S., Pardi, R.R., Marcus, L.F., and Sperling, J.A., 1980c, The determination of the magnitude and date of dip-slip faulting by discordance in sets of sea-level curves, USGS Open File Report 80-842, p. 156-158.

Newman, W.S., Marcus, L.F., Pardi, R.R., and Vulis, I., 1984, The trouble with sea level: 4 pp. abstract in: International Symposium on Late Quaternary Sea Level and Coastal Evolution, Mar del Plata, Argentina.

Newman, W.S., Cinquemani, L.J., Sperling, J.A., Marcus, L., and Pardi, R.R., 1987, Holocene neotectonics and the Ramapo Fault Zone sea-level anomaly: a study of varying marine transgression rates in the lower Hudson estuary, New York and New Jersey, p. 97-114 in: D. Nummedal, O. H. Pilkey, and J.D. Howard, eds., Sea-Level Fluctuations and Coastal Evolution, SEPM Special Publication No. 41, Tulsa, Oklahoma.

Oldale, R.N., and O'Hara, C.J., 1980, New radiocarbon dates from the inner shelf off southeastern Massachusetts and a local sea-level rise curve for the past 12,000 years: Geology, v. 8, p. 102-106.

Officer, C.B., Newman, W.S., Sullivan, J.M., and Lynch, D.R., 1987, Glacial isostatic adjustment and mantle viscosity: Journal of Geophysical Research, in press.

Pardi, R.R., and Newman, W.S., 1987, Late Quaternary sea levels along the Atlantic coast of North America: Journal of Coastal Research, v. 3, no. 3, p. 325-330.

Peltier, W.R., 1974, The impulse response of a Maxwell earth: Reviews of Geophysics and Space Physics, v. 12, p. 649-699.

Peltier, W.R., 1980, Ice sheets, oceans, and the earth's shape: p. 45-63 in: N.-A. Morner, ed., Earth Rheology, Isostasy and Eustasy, Wiley, New York, 599 pp.

Petersen, K.S., 1985, The Late Quaternary history of Denmark: The Weichselian icesheets and land/sea configuration in the Late Pleistocene and Holocene: Journal of Danish Archaeology, v. 4, p. 7-22.

Quinlan, G., and Beaumont, C., 1981, A comparison of observed and theoretical relative sea levels in Atlantic Canada: Canadian Journal of Earth Sciences, v. 18, p. 1146-1163.

Redfield, A.C., 1967, Postglacial change in sea level in the western North Atlantic Ocean: Science, v. 157, p. 687-692.

Redfield, A.C., and Rubin, M., 1962, The age of salt marsh peat and its relation to recent changes in sea level at Barnstable, Massachusetts: National Academy Science Proceedings, v. 48, p. 1728-1735.

Scholl, D.W., 1964, Recent sedimentary record in mangrove swamps and rise in sea level over the southwestern coast of Florida: Part I: Marine Geology, v. 1, p. 344-366.

Scholl, D.W., and Stuiver, M., 1967a, Recent submergence of southern Florida: A comparison with adjacent coasts: Geological Society of America Bulletin, v. 78, p. 437-454.

Scholl, D.W., and Stuiver, M., 1967b, Recent submergence of southern Florida: a Reply: Geological Society of America Bulletin, v. 78, p. 1195-1198.

Scott, D.B., and Medioli, F.S., 1982, Micropaleontological documentation for early Holocene fall of relative sea level on the Atlantic coast of Nova Scotia: Geology, v. 10, no. 5, p. 278-281.

Shepard, F.P., 1963, Thirty-five thousand years of sea level: p. 1-10 in: T. Clements, ed., Essays in Marine Geology in Honor of K.O. Emery, University of Southern California Press, Los Angeles, 201 pp.

Taira, K., 1975, Holocene crustal movements in Taiwan as identified by radiocarbon dating of marine fossils and driftwood: Tectonophysics, v. 28, p. T1-T5.

Van Andel, T.H., and Laborel, J., 1964, Recent high relative sea-level stand near Recife, Brazil: Science, v. 145, p. 580-581.

Vita-Finzi, C., 1975, Quaternary deposits in the Iranian Makran: Geographical Journal, v. 141, p. 515-520.

Walcott, R.I., 1972a, Late Quaternary vertical movements in eastern North America: quantitative evidence of glacio-isostatic rebound: Reviews of Geophysics and Space Physics, v. 10, p. 849-884.

Walcott, R.I., 1972b, Past sea levels, eustasy, and deformation of the earth: Quaternary Research, v. 2, p. 1-14.

LIST OF REVIEWERS

Dr. Daniel F. Belknap
University of Maine

Dr. P. Bellwood
Australian National University

Dr. B. Carter
The University of Ulster

Dr. Rhodes W. Fairbridge
Lamont-Doherty Geological Observatory

Dr. H. Faure
CNRS - University Luminy

Dr. Claude Hillaire-Marcel
Universite du Quebec a Montreal

Dr. David Hopley
James Cook University

Dr. Saskia Jelgersma
Geological Survey of the Netherlands

Francine McCarthy
Dalhousie University

Dr. Paolo A. Pirazzoli
CNRS - INTERGEO

Dr. B.R. Pelletier
Geological Survey of Canada

Dr. Wylie Poag
U.S. Geological Survey

Dr. David B. Scott
Dalhousie University

Dr. Ian Shennan
University of Durham

Dr. D.E. Smith
Coventry Lanchester
            Polytechnic

Dr. C.E. Stearns
Tufts University

Dr. John R. Suter
Louisiana Geological Survey

Dr. Keith R. Thompson
Dalhousie University

Dr. P.L. Woodworth
Bidston Observatory